AF544968

Reviews of Environmental Contamination and Toxicology

VOLUME 245

More information about this series at http://www.springer.com/series/398

Reviews of Environmental Contamination and Toxicology

Editor
Pim de Voogt

Volume 245

ISSN 0179-5953 ISSN 2197-6554 (electronic)
Reviews of Environmental Contamination and Toxicology
ISBN 978-3-319-75036-1 ISBN 978-3-319-75037-8 (eBook)
DOI 10.1007/978-3-319-75037-8

Printed on acid-free paper

This Springer imprint is published by Springer Nature
The registered company is Springer International Publishing AG
The registered company address is: Gewerbestrasse 11, 6330 Cham, Switzerland

Foreword

International concern in scientific, industrial, and governmental communities over traces of xenobiotics in foods and in both abiotic and biotic environments has justified the present triumvirate of specialized publications in this field: comprehensive reviews, rapidly published research papers and progress reports, and archival documentations These three international publications are integrated and scheduled to provide the coherency essential for nonduplicative and current progress in a field as dynamic and complex as environmental contamination and toxicology. This series is reserved exclusively for the diversified literature on "toxic" chemicals in our food, our feeds, our homes, recreational and working surroundings, our domestic animals, our wildlife, and ourselves. Tremendous efforts worldwide have been mobilized to evaluate the nature, presence, magnitude, fate, and toxicology of the chemicals loosed upon the Earth. Among the sequelae of this broad new emphasis is an undeniable need for an articulated set of authoritative publications, where one can find the latest important world literature produced by these emerging areas of science together with documentation of pertinent ancillary legislation.

Research directors and legislative or administrative advisers do not have the time to scan the escalating number of technical publications that may contain articles important to current responsibility. Rather, these individuals need the background provided by detailed reviews and the assurance that the latest information is made available to them, all with minimal literature searching. Similarly, the scientist assigned or attracted to a new problem is required to glean all literature pertinent to the task, to publish new developments or important new experimental details quickly, to inform others of findings that might alter their own efforts, and eventually to publish all his/her supporting data and conclusions for archival purposes.

In the fields of environmental contamination and toxicology, the sum of these concerns and responsibilities is decisively addressed by the uniform, encompassing, and timely publication format of the Springer triumvirate:

Reviews of Environmental Contamination and Toxicology [Vol. 1 through 97 (1962–1986) as Residue Reviews] for detailed review articles concerned with any aspects of chemical contaminants, including pesticides, in the total environment with toxicological considerations and consequences.

Bulletin of Environmental Contamination and Toxicology (Vol. 1 in 1966) for rapid publication of short reports of significant advances and discoveries in the fields of air, soil, water, and food contamination and pollution as well as methodology and other disciplines concerned with the introduction, presence, and effects of toxicants in the total environment.

Archives of Environmental Contamination and Toxicology (Vol. 1 in 1973) for important complete articles emphasizing and describing original experimental or theoretical research work pertaining to the scientific aspects of chemical contaminants in the environment.

The individual editors of these three publications comprise the joint Coordinating Board of Editors with referral within the board of manuscripts submitted to one publication but deemed by major emphasis or length more suitable for one of the others.

Coordinating Board of Editors

Preface

The role of *Reviews* is to publish detailed scientific review articles on all aspects of environmental contamination and associated (eco)toxicological consequences. Such articles facilitate the often complex task of accessing and interpreting cogent scientific data within the confines of one or more closely related research fields.

In the 50+ years since *Reviews of Environmental Contamination and Toxicology* (formerly *Residue Reviews)* was first published, the number, scope, and complexity of environmental pollution incidents have grown unabated. During this entire period, the emphasis has been on publishing articles that address the presence and toxicity of environmental contaminants. New research is published each year on a myriad of environmental pollution issues facing people worldwide. This fact, and the routine discovery and reporting of emerging contaminants and new environmental contamination cases, creates an increasingly important function for *Reviews*. The staggering volume of scientific literature demands remedy by which data can be synthesized and made available to readers in an abridged form. *Reviews* addresses this need and provides detailed reviews worldwide to key scientists and science or policy administrators, whether employed by government, universities, nongovernmental organizations, or the private sector.

There is a panoply of environmental issues and concerns on which many scientists have focused their research in past years. The scope of this list is quite broad, encompassing environmental events globally that affect marine and terrestrial ecosystems; biotic and abiotic environments; impacts on plants, humans, and wildlife; and pollutants, both chemical and radioactive; as well as the ravages of environmental disease in virtually all environmental media (soil, water, air). New or enhanced safety and environmental concerns have emerged in the last decade to be added to incidents covered by the media, studied by scientists, and addressed by governmental and private institutions. Among these are events so striking that they are creating a paradigm shift. Two in particular are at the center of ever increasing media as well as scientific attention: bioterrorism and global warming. Unfortunately, these very worrisome issues are now superimposed on the already extensive list of ongoing environmental challenges.

The ultimate role of publishing scientific environmental research is to enhance understanding of the environment in ways that allow the public to be better informed or, in other words, to enable the public to have access to sufficient information. Because the public gets most of its information on science and technology from internet, TV news, and reports, the role for scientists as interpreters and brokers of scientific information to the public will grow rather than diminish. Environmentalism is an important global political force, resulting in the emergence of multinational consortia to control pollution and the evolution of the environmental ethic. Will the new politics of the twenty-first century involve a consortium of technologists and environmentalists, or a progressive confrontation? These matters are of genuine concern to governmental agencies and legislative bodies around the world.

For those who make the decisions about how our planet is managed, there is an ongoing need for continual surveillance and intelligent controls to avoid endangering the environment, public health, and wildlife. Ensuring safety-in-use of the many chemicals involved in our highly industrialized culture is a dynamic challenge, because the old, established materials are continually being displaced by newly developed molecules more acceptable to federal and state regulatory agencies, public health officials, and environmentalists. New legislation that will deal in an appropriate manner with this challenge is currently in the making or has been implemented recently, such as the REACH legislation in Europe. These regulations demand scientifically sound and documented dossiers on new chemicals.

Reviews publishes synoptic articles designed to treat the presence, fate, and, if possible, the safety of xenobiotics in any segment of the environment. These reviews can be either general or specific, but properly lie in the domains of analytical chemistry and its methodology, biochemistry, human and animal medicine, legislation, pharmacology, physiology, (eco)toxicology, and regulation. Certain affairs in food technology concerned specifically with pesticide and other food-additive problems may also be appropriate.

Because manuscripts are published in the order in which they are received in final form, it may seem that some important aspects have been neglected at times. However, these apparent omissions are recognized, and pertinent manuscripts are likely in preparation or planned. The field is so very large and the interests in it are so varied that the editor and the editorial board earnestly solicit authors and suggestions of underrepresented topics to make this international book series yet more useful and worthwhile.

Justification for the preparation of any review for this book series is that it deals with some aspect of the many real problems arising from the presence of anthropogenic chemicals in our surroundings. Thus, manuscripts may encompass case studies from any country. Additionally, chemical contamination in any manner of air, water, soil, or plant or animal life is within these objectives and their scope.

Manuscripts are often contributed by invitation. However, nominations for new topics or topics in areas that are rapidly advancing are welcome. Preliminary communication with the Editor-in-Chief is recommended before volunteered review manuscripts are submitted. *Reviews* is registered in WebofScience™.

Inclusion in the Science Citation Index serves to encourage scientists in academia to contribute to the series. The impact factor in recent years has increased from 2.5 in 2009 to almost 4 in 2013. The Editor-in-Chief and the Editorial Board strive for a further increase of the journal impact factor by actively inviting authors to submit manuscripts.

Amsterdam, The Netherlands Pim de Voogt
January 2015

Contents

Contributors

Andiranel Banegas Department of Aquatic Systems, Faculty of Environmental Sciences, EULA-Chile Centre, Universidad de Concepción, Concepción, Chile

Department of Sciences Biology Unit, Danlí Technological Campus, Universidad Nacional Autónoma de Honduras, Danlí, Honduras

Ricardo Barra Department of Aquatic Systems, Faculty of Environmental Sciences, EULA-Chile Centre, Universidad de Concepción, Concepción, Chile

Ron Biever Smithers Viscient, Wareham, MA, USA

Martin Brtnicky Department of Geology and Pedology, Faculty of Forestry and Wood Technology, Mendel University in Brno, Brno, Czech Republic

Central European Institute of Technology, Brno University of Technology, Brno, Czech Republic

José E. Celis Department of Animal Science, Faculty of Veterinary Sciences, Universidad de Concepción, Chillán, Chile

Gustavo Chiang Melimoyu Ecosystem Research Institute, Santiago, Chile

Hana Cihlarova Department of Geology and Pedology, Faculty of Forestry and Wood Technology, Mendel University in Brno, Brno, Czech Republic

Winfred Espejo Department of Aquatic Systems, Faculty of Environmental Sciences, EULA-Chile Centre, Universidad de Concepción, Concepción, Chile

Radka Fryzova Department of Geology and Pedology, Faculty of Forestry and Wood Technology, Mendel University in Brno, Brno, Czech Republic

Central European Institute of Technology, Brno University of Technology, Brno, Czech Republic

Robert M. Gogal, Jr. Department of Veterinary Biosciences and Diagnostic Imagining, College of Veterinary Medicine, University of Georgia, Athens, GA, USA

Daniel González-Acuña Department of Animal Science, Faculty of Veterinary Sciences, Universidad de Concepción, Chillán, Chile

Patrick D. Guiney University of Wisconsin-Madison, Madison, WI, USA

Jan Hladky Department of Geology and Pedology, Faculty of Forestry and Wood Technology, Mendel University in Brno, Brno, Czech Republic

Central European Institute of Technology, Mendel University in Brno, Brno, Czech Republic

Steven D. Holladay Department of Veterinary Biosciences and Diagnostic Imagining, College of Veterinary Medicine, University of Georgia, Athens, GA, USA

Natalie Karouna-Renier USGS Patuxent Wildlife Research Center, Beltsville, MD, USA

Ioanna Katsiadaki Centre for Environment, Fisheries and Aquaculture Science, Weymouth, Dorset, UK

Jindrich Kynicky Department of Geology and Pedology, Faculty of Forestry and Wood Technology, Mendel University in Brno, Brno, Czech Republic

Central European Institute of Technology, Brno University of Technology, Brno, Czech Republic

Laurent Lagadic Bayer AG, Research and Development, Crop Science Division, Environmental Safety, Monheim am Rhein, Germany

Pavla Martinkova Central European Institute of Technology, Brno University of Technology, Brno, Czech Republic

Faculty of Military Health Sciences, University of Defence, Hradec Kralove, Czech Republic

James P. Meador Environmental and Fisheries Sciences Division, Northwest Fisheries Science Center, National Marine Fisheries Service, National Oceanic and Atmospheric Administration, Seattle, WA, USA

Miroslav Pohanka Department of Geology and Pedology, Faculty of Forestry and Wood Technology, Mendel University in Brno, Brno, Czech Republic

Faculty of Military Health Sciences, University of Defence, Hradec Kralove, Czech Republic

Tamar Schwarz Centre for Environment, Fisheries and Aquaculture Science, Weymouth, Dorset, UK

Robert J. Williams Department of Veterinary Biosciences and Diagnostic Imagining, College of Veterinary Medicine, University of Georgia, Athens, GA, USA

Susan M. Williams Poultry Diagnostic and Research Center, College of Veterinary Medicine, University of Georgia, Athens, GA, USA

A Global Overview of Exposure Levels and Biological Effects of Trace Elements in Penguins

Winfred Espejo, José E. Celis, Daniel González-Acuña, Andiranel Banegas, Ricardo Barra, and Gustavo Chiang

Contents

W. Espejo • R. Barra
Department of Aquatic Systems, Faculty of Environmental Sciences, EULA-Chile Centre, Universidad de Concepción, P.O. Box 160-C, Concepción, Chile
e-mail: winfredespejo@udec.cl; ricbarra@udec.cl

J.E. Celis (✉) • D. González-Acuña
Department of Animal Science, Faculty of Veterinary Sciences, Universidad de Concepción, P.O. Box 537, Chillán, Chile
e-mail: jcelis@udec.cl; danigonz@udec.cl

A. Banegas
Department of Aquatic Systems, Faculty of Environmental Sciences, EULA-Chile Centre, Universidad de Concepción, P.O. Box 160-C, Concepción, Chile

Department of Sciences Biology Unit, Danlí Technological Campus, Universidad Nacional Autónoma de Honduras, Danlí, Honduras
e-mail: bbanegas@udec.cl

G. Chiang
Melimoyu Ecosystem Research Institute, Santiago, Chile
e-mail: gchiang@fundacionmeri.cl

P. de Voogt (ed.), *Reviews of Environmental Contamination and Toxicology Volume 245*, Reviews of Environmental Contamination and Toxicology 245,
DOI 10.1007/398_2017_5

Acronyms

Al	Aluminum
As	Arsenic
Ca	Calcium
Cd	Cadmium
Co	Cobalt
Cr	Chromium
Cu	Copper
Fe	Iron
Hg	Mercury
Mn	Manganese
Ni	Nickel
Pb	Lead
Se	Selenium
V	Vanadium
Zn	Zinc

Highlights

- Most of the studies of metals in penguins have been carried out in Antarctic and subantarctic islands. However, there is a lack of data from lower latitudes where other important penguin species inhabit.
- The levels of metals are reported mainly in feathers and excreta. Further research in other biological matrices such as internal organs and blood is required.
- Further research in the issue of biological effects caused by metals is needed.
- Little is known about the interaction among the metals which could activate certain mechanisms of detoxification of the body of the penguins.

1 Introduction

Trace metal toxicity is one of the major stressors leading to hazardous effects on biota (Bargagli 2001; Zhang and Ma 2011). In aquatic environments, trace element contamination is a great concern, due to the implications these chemicals may have on both wildlife and human health (Lavoie et al. 2013; Prashanth et al. 2016). These elements enter the water through natural erosion, geochemical cycles, industrial processes, and agricultural practices (Burger and Gochfeld 2000a). In birds, some metals can produce severe adverse effects such as difficulty in flying, walking and standing, paralysis, and an increase in mortality (Newman 2015). In order to monitor the occurrence of environmental pollutants in marine ecosystems, the use of aquatic birds has greatly increased, because they can accumulate trace elements in diverse tissues, such as eggs, feathers, or liver, thus can be used to indirectly evaluate in a proper way the toxicological status of the marine ecosystem under study (Savinov et al. 2003). Moreover, seabird diet and feeding ecology can differ in response to climate change, thus affecting exposure to metals over time (Braune et al. 2014). Evidence indicates that the concentrations of certain pollutants in seabirds have a lower variation coefficient than that observed in fishes or marine mammals, so that the analysis of a relatively low number of samples of birds is similar to that obtained by analyzing a significantly higher number in other groups of animals (Pérez-López et al. 2005). Birds tend to be more sensitive to environmental contaminants than other vertebrates (Zhang and Ma 2011), thus ecotoxicological studies on seabirds have proliferated in recent years (Casini et al. 2001; Barbieri et al. 2010; Barbosa et al. 2013; Celis et al. 2014; Kehrig et al. 2015).

The study of trace elements in penguins is valuable, because they are animals that exclusively inhabit the Southern Hemisphere and represent about 90% of the bird biomass of the Southern Ocean (Williams 1990). Penguins are present in different systems in the Antarctic, subantarctic islands of the Pacific, Atlantic, and Indian oceans, as well as on the coasts of Australia, South Africa, South America, and the Galapagos (García and Boersma 2013). Penguins are useful indicators of the degree of contamination by trace elements in the environment, because they are highly specialized animals that swim and dive in search for food, are widely distributed, and are organisms usually found at the top of the trophic web (De Moreno et al. 1997; Boersma 2008; Fig. 1). Additionally, penguins are extremely interesting as bioindicators because of their intense molting process (Carravieri et al. 2014), and because they can be finicky eaters with a restricted diet (Lescroël et al. 2004; Jerez et al. 2011).

The different penguin species (order Sphenisciformes, family Spheniscidae) can be classified in the genera *Aptenodytes*, *Eudyptes*, *Eudyptula*, *Megadyptes*, *Pygoscelis,* and *Spheniscus*. These species have in common the fact of presenting serious risks of survival in the future, because about two thirds of penguin species are on the Red List of Threatened Species of the International Union for Conservation of Nature (UICN 2016). Contamination, climate change, fishing, alterations of ecosystems, diseases, and even tourism are their major threats (García and Boersma 2013). The lack of knowledge about the effects of trace elements in seabirds is a main threat to their population sustainability (Sanchez-Hernandez 2000).

The study of the biological effects of toxic trace elements ingested by penguins is of great relevance because it may contribute knowledge of possible consequences in nature (Nordberg and Nordberg 2016). Moreover, evidence has revealed that some trace elements such as As, Cd, Hg, Mn, Pb, and Zn can affect the endocrine system of animals and humans, producing alterations in physiological functions (Iavicoli et al. 2009). Given the wide distribution of penguins, data concerning trace element concentrations in different biotic matrices of penguins are summarized

Fig. 1 Sources of trace elements in the environment, bioaccumulation, biomagnification, and effects on penguins. According to Newman (2015), bioaccumulation is the net accumulation of a contaminant on an organism from all sources including water, air, and solid phases (food, soil, sediment, or fine particles suspended in air or water) in the environment. Biomonitoring is the use of organisms to monitor contamination and its possible effects on biota (at individual level, population, or communities) and ecosystems

here to be used as a first background database for contamination detection in marine ecosystems.

2 Materials and Methods

In order to identify the interaction of penguin species and trace elements, a systematic study of the existing literature on the concentrations of trace elements in different biotic matrices studied was conducted. These biological matrices correspond to guano, feathers, eggs, blood, stomach contents, and internal organs. By using databases such as Direct, Springer, Scopus, and Web of Science, different keywords were used. Among them, "trace element," "heavy metal," "trace metals," "mercury," "aluminum," "arsenic," "cadmium," "lead," "zinc," "copper," "pollution," "persistent pollutants," "monitoring," "biomonitoring," "penguin," "seabirds," "eggs," "blood," "guano," "droppings," "feathers," "tissues," "organs," and "Antarctic" can be mentioned. Furthermore, the list of references of each publication was reviewed to identify additional documents on the issue not previously found. Selection criteria strictly corresponded to trace element concentrations on the basis of dry weight (dw) based on studies performed in situ.

Subsequently this information was summarized in tables. Mean levels of exposure to trace elements in penguins were compared with other marine, aquatic, or terrestrial birds from different parts of the world. In addition, maps of the distribution of penguins were included along with records of trace elements, based on the identification of the different colonies of penguins worldwide (Boersma 2008). Information about the investigations related to the exposure and effect of trace elements in penguins was also considered.

A bubble chart was built upon mean concentrations of each trace element reported in gentoo penguins (*Pygoscelis papua*) at South Shetland Islands to see similarities and differences according to biological matrices. First, concentration values were normalized by log $(x + 1)$ to remove any weighting from dominant peaks and then analyzed with a Bray-Curtis similarity matrix (Clarke et al. 2006). Finally, the resultant similarity matrix was analyzed in a two-dimensional multidimensional scaling (MDS) plot.

Values of the trophic transference coefficient (TTC), as the ratio between the level of a certain element in the penguin's body (liver, kidneys, bones, and muscles together) and the level of the element in stomach contents (Suedel et al. 1994), were calculated taking into consideration the mean concentration of each element.

3 Exposure to Trace Elements and Its Effects on Penguins

Trace element concentrations measured in different biotic matrices of different species of penguins are presented in Tables 1, 2, 3, 4, 5, 6, 7, 8, and 9. The most commonly reported trace elements in penguins are Al, As, Cd, Cu, Hg, Mn, Pb, and Zn, whereas the gentoo penguin is the species that displays the highest concentration of most trace elements studied. Other trace elements such as Co, Cr, Fe, Ni, and Se have been poorly studied (Szopińska et al. 2016). Levels of Fe (23.37–164.26 μg/g) have been reported in feathers of pygoscelid penguins from Antarctica (Metcheva et al. 2006; Jerez et al. 2011). In the same matrix and region, levels of Se (1.8–2.0 μg/g) have been linked with a major exposure to Cd and Hg (Jerez et al. 2011), since Se is known to have a detoxifying effect of these metals (Smichowski et al. 2006). Cobalt levels have been reported in feathers of chinstrap and gentoo penguins from Antarctica (0.17–0.25 μg/g, Metcheva et al. 2006). Nickel has been reported in chinstrap penguins from Antarctica in feces (3.2–3.7 μg/g), liver (0.07 μg/g), and muscle (<0.03 μg/g) by Metcheva et al. (2006), and in feathers of pygoscelid penguins (0.24–1.18 μg/g, Jerez et al. 2011). In the organs of most avian wildlife species from unpolluted ecosystems, Ni concentrations may vary greatly (0.1–2.0 μg/g, Outridge and Scheuhammer 1993). Chromium levels (1.15–8.08 μg/g) were reported by Jerez et al. (2011) in

Table 1 Mean trace metal levels (µg/g, dw) in feathers of different penguin species worldwide

Species	N	Al	As	Cd	Pb	Hg	Cu	Zn	Mn	Locations	Date[a]	References
P. adeliae	3	–	–	–	–	0.82±0.13	–	–	–	Terra Nova Bay[b]	1989–1990	Bargagli et al. (1998)
	1	–	–	–	–	1.40	–	61.5	–	Admiralty Bay[c]	2004	Santos et al. (2006)
	n/i	–	–	–	1.50	0.60	–	–	–	Zhongshan Station[d]	2001	Yin et al. (2008)
	1	3.56	0.04	0.12	<0.01	–	16.21	70.41	0.21	King George is.[c]	2007–2010	Jerez et al. (2013a)
	2	0.71±0.43	0.06±0.001	0.08±0.01	0.06±0.09	–	16.22±0.51	60.59±2.02	<0.01	Avian is.[e]	2007–2010	Jerez et al. (2013a)
	1[f]	52.44	0.08	0.01	<0.01	–	19.29	83.90	1.15	King George is.[c]	2007–2010	Jerez et al. (2013a)
	5	64.3±61.75	0.17±0.11	0.13±0.08	0.24±0.38	–	13.32±8.22	61.11±20.3	2.01±0.52	King George is.[c]	2008–2009	Jerez et al. (2013b)
	4	–	–	0.30	0.50	0.90	–	–	–	Edmonson Point[b]	1995	Ancora et al. (2002)
	25	43.36±69.03	–	–	0.64±1.09	–	12.68±7.09	50.84±17.38	1.30±1.16	King George is.[c]	2005–2007	Jerez et al. (2011)
	21	8.62±6.41	0.07±0.04	0.04±0.05	0.32±0.36	–	13.41±2.6	82.45±13.1	1.16±1.26	Yalour is.[e]	2005–2007	Jerez et al. (2011)
	22	5.08±3.03	0.07±0.03	0.04±0.02	0.14±0.21	–	13.16±3.04	77.69±15.17	0.34±0.49	Avian is.[e]	2005–2007	Jerez et al. (2011)
	10	–	–	–	–	0.66±0.2	–	–	–	Adélie land[g]	2007	Carravieri et al. (2016)

	12	–	–	–	–	0.43±0.13	–	–	–	Adélie land[g]	2006	Carravieri et al. (2016)
	10[f]	–	–	–	–	0.19±0.06	–	–	–	Adélie land[g]	2007	Carravieri et al. (2016)
P. antarctica	25	132.4±198.09	0.10	–	1.76±1.74	–	20.29±8.3	77.12±45.15	1.66±0.98	King George is.[c]	2007	Jerez et al. (2011)
	10	26.07±9.97	0.01±0.0001	0.04±0.03	0.15±0.12	–	14.93±6.1	72.21±28.96	0.92±0.59	Livingston is.[c]	2000	Jerez et al. (2011)
	25	203.13±194.65	0.10±0.1	0.08±0.04	0.32±0.22	–	16.39±3.44	82.40±15.64	3.26±2.68	Deception is.[c]	2005–2007	Jerez et al. (2011)
	20	14.26±9.72	0.05±0.03	0.10±0.05	0.14±0.09	–	19.23±3.65	97.27±21.35	0.29±0.39	Ronge is.[e]	2005–2007	Jerez et al. (2011)
	3	8.99±7.86	0.07±0.01	0.31±0.22	0.81±0.84	–	19.60±1.7	62.29±20.01	0.05±0.03	Deception is.[c]	2005–2007	Jerez et al. (2013a)
	2	7.99±10.02	0.31±0.21	0.01±0.01	0.02±0.03	–	15.29±0.34	94.75±2.37	0.21±0.30	King George is.[c]	2005–2007	Jerez et al. (2013a)
	5	142±206.33	0.48±0.3	0.02±0.03	0.06±0.04	–	18.57±2.78	94.99±5.29	2.25±3.17	Deception is.[c]	2007–2010	Jerez et al. (2013b)
	29	26±8	0.45±0.2	0.30±0.07	1.73±0.94	–	18.5±3	87.0±5	1.5±0.7	Livingston is.[c]	2007–2010	Metcheva et al. (2006)
	32	26±11	2.4[a]	0.2[a]	1.66±1.2	–	18±2	75±7	1.6±0.43	Livingston is.[c]	2008–2009	Metcheva et al. (2006)
P. papua	1	–	–	–	–	0.54	–	90.70	–	Admiralty Bay[c]	2002	Santos et al. (2006)
	14	40±10	0.88±0.32	0.21±0.06	1.7±1.3	–	17±4	106±8	1.5±0.73	Livingston is.[c]	2002	Metcheva et al. (2006)

(continued)

Table 1 (continued)

Species	N	Al	As	Cd	Pb	Hg	Cu	Zn	Mn	Locations	Date[a]	References
	28	46±22	(<0.6–4.0)[h]	(0.15–0.43)[h]	1.57±1.13	–	16±2	89±7	2.6±0.99	Livingston is.[c]	2003	Metcheva et al. (2006)
	1[i]	–	–	0.50±0.12	<1.90	–	15.9±0.12	73.0±0.54	1.65±0.15	Livingston is.[c]	2002–2006	Metcheva et al. (2010)
	14	37.0±6.9	(0.3–0.69)[h]	(0.05–0.41)[h]	1.52±0.5	–	17.0±3.1	92.0±4.6	1.70±0.2	Livingston is.[c]	2006–2007	Metcheva et al. (2011)
	20	39.76±24.74	0.05±0.04	0.03±0.03	0.51±0.46	–	16.44±3.16	85.12±14.84	1.80±1.28	King George is.[c]	2005–2007	Jerez et al. (2011)
	14	40.17±42.37	0.02±0.02	0.02±0.01	0.17±0.23	–	13.88±2.89	75.41±15.85	1.17±1.05	Livingston is.[c]	2005–2007	Jerez et al. (2011)
	17	22.92±27.87	0.04±0.02	0.03±0.01	0.25±0.44	–	16.02±2.09	72.89±7.46	0.46±0.58	Ronge is.[e]	2005–2007	Jerez et al. (2011)
	8	17.19±11.9	0.07	0.03±0.04	0.31±0.09	–	13.42±5.09	61.71±18.01	0.68±0.66	Paradise Bay[e]	2005–2007	Jerez et al. (2011)
	11	–	–	–	–	5.90±1.91[j]	–	–	–	Crozet is.[k]	2007	Carravieri et al. (2016)
	12	–	–	–	–	5.23±1.12	–	–	–	Crozet is.[k]	2006	Carravieri et al. (2016)
	12[l]	–	–	–	–	1.88±0.46	–	–	–	Crozet is.[k]	2007	Carravieri et al. (2016)
	4[f]	15.72±19.24	0.09±0.07	0.02±0.01	0.07±0.13	–	16.02±5.40	119.72±21.81	0.27±0.35	King George is.[c]	2007–2010	Jerez et al. (2013a)
	2	6.71±5.15	0.10±0.05	0.02±0.004	0.33±0.31	–	19.26±0.94	69.49±6.32	0.06±0.09	King George is.[c]	2007–2010	Jerez et al. (2013a)

	5	68.55±76.39	0.12±0.05	0.06±0.04	0.87±0.86	–	6.87±1.54	80.59±10.85	0.95±0.69	King George is.[c]	2008–2009	Jerez et al. (2013b)
	10	–	–	0.05±0.07	0.06±0.04	–	13.74±1.81	36.89±6.26	0.75±0.28	Neko Harbor[e]	2014	Celis et al. (2015b)
	10	–	–	0.09±0.07	0.10±0.17	–	14.98±4.09	33.26±4.04	0.27±0.37	Doumer is.[e]	2014	Celis et al. (2015b)
	10	–	–	0.14±0.09	0.60±0.34	–	19.65±2.25	64.19±10.67	1.23±0.46	Stranger Point[c]	2014	Celis et al. (2015b)
	10	–	–	0.21±0.28	0.63±0.27	–	20.89±4.3	64.07±10.73	1.19±0.68	O'Higgins Base[e]	2014	Celis et al. (2015b)
	55	–	–	–	–	0.97±0.67	–	–	–	South Georgia[d]	2009	Pedro et al. (2015)
	29	–	–	–	–	1.1±0.62	–	–	–	South Georgia[d]	2011	Pedro et al. (2015)
	12	–	–	–	–	5.85±3.00	–	–	–	Kerguelen is.[k]	2007	Carravieri et al. (2013)
	12	–	–	–	–	4.96±2.44	–	–	–	Kerguelen is.[k]	2006	Carravieri et al. (2013)
	12[l]	–	–	–	–	2.45±0.67	–	–	–	Kerguelen is.[k]	2007	Carravieri et al. (2013)
	12	–	–	–	–	1.44±0.44	–	–	–	Kerguelen is.[k]	2007	Carravieri et al. (2013)
A. forsteri	3	–	–	–	–	0.98±0.21	–	–	–	Terra Nova Bay[b]	1989–1990	Bargagli et al. (1998)
	17	–	–	–	–	1.77±0.37	–	–	–	Adélie Land[g]	2007	Carravieri et al. (2016)

(continued)

Table 1 (continued)

Species	N	Al	As	Cd	Pb	Hg	Cu	Zn	Mn	Locations	Date[a]	References
	12[f]	–	–	–	–	0.61±0.11	–	–	–	Adélie Land[g]	2007	Carravieri et al. (2016)
A. patagonicus	31	–	–	–	–	1.98±0.73	–	–	–	Crozet is.[k]	2000–2001	Scheifler et al. (2005)
	10	–	–	–	–	2.66±0.86	–	–	–	Crozet is.[k]	1966–1974	Scheifler et al. (2005)
	12	–	–	–	–	2.22±0.59	–	–	–	Kerguelen is.[k]	2007	Carravieri et al. (2013)
	12	–	–	–	–	2.17±0.52	–	–	–	Kerguelen is.[k]	2006	Carravieri et al. (2013)
	12[f]	–	–	–	–	1.79±0.55	–	–	–	Kerguelen is.[k]	2007	Carravieri et al. (2013)
	12[l]	–	–	–	–	1.12±0.16	–	–	–	Kerguelen is.[k]	2007	Carravieri et al. (2013)
	12	–	–	–	–	2.94±0.47	–	–	–	Crozet is.[k]	2006	Carravieri et al. (2016)
	12	–	–	–	–	2.89±0.73	–	–	–	Crozet is.[k]	2007	Carravieri et al. (2016)
	12[f]	–	–	–	–	1.80±0.24	–	–	–	Crozet is.[k]	2007	Carravieri et al. (2016)

E. chrysolophus	12	–	–	–	–	2.24±0.29	–	–	–	Kerguelen is.[k]	2007	Carravieri et al. (2013)
	12	–	–	–	–	2.08±0.35	–	–	–	Kerguelen is.[k]	2006	Carravieri et al. (2013)
	12[l]	–	–	–	–	0.36±0.07	–	–	–	Kerguelen is.[k]	2007	Carravieri et al. (2013)
	12	–	–	–	–	2.48±0.35	–	–	–	Crozet is.[k]	2007	Carravieri et al. (2016)
	12	–	–	–	–	2.09±0.31	–	–	–	Crozet is.[k]	2006	Carravieri et al. (2016)
	12[f]	–	–	–	–	0.43±0.10	–	–	–	Crozet is.[k]	2007	Carravieri et al. (2016)
S. magellanicus	21	–	–	–	–	0.206±0.098	–	–	–	Punta Tombo [m]	2007	Frias et al. (2012)
	18[n]	–	–	–	–	0.123±102	–	–	–	Punta Tombo [m]	2007	Frias et al. (2012)
	37[f]	–	–	–	–	0.033±0.052	–	–	–	Punta Tombo [m]	2007	Frias et al. (2012)
	22	–	–	0.13±0.07	0.14±0.08	0.78±0.44	–	–	–	Rio Grande do Sul[o]	n/i	Kehrig et al. (2015)
E. chrysocome	12	–	–	–	–	1.96±0.41	–	–	–	Kerguelen is.[k]	2007	Carravieri et al. (2013)
	12	–	–	–	–	1.92±0.35	–	–	–	Kerguelen is.[k]	2006	Carravieri et al. (2013)

(continued)

Table 1 (continued)

Species	N	Al	As	Cd	Pb	Hg	Cu	Zn	Mn	Locations	Date[a]	References
	12[f]	–	–	–	–	0.27±0.06	–	–	–	Kerguelen is.[k]	2007	Carravieri et al. (2013)
	12	–	–	–	–	1.79±0.37	–	–	–	Crozet is.[k]	2007	Carravieri et al. (2016)
	12	–	–	–	–	1.62±0.35	–	–	–	Crozet is.[k]	2006	Carravieri et al. (2016)
	12[f]	–	–	–	–	0.34±0.05	–	–	–	Crozet is.[k]	2007	Carravieri et al. (2016)
E. minor	13	40.38±22.96	0.16±0.05	0.04±0.02	0.42±0.20	4.13±0.98	11.42±2.19	84.77±11.28	–	St. Kilda [p]	2012	Finger et al. (2015)
	12	16.78±16.11	0.18±0.1	0.04±0.03	0.08±0.03	2.70±0.37	10.77±1.5	80.58±8.06	–	Phillip is.[p]	2012	Finger et al. (2015)
	10	6.25±2.79	0.09±0.03	0.06±0.02	0.10±0.05	1.50±0.82	10.54±2.07	76.80±7.15	–	Notch is.[p]	2012	Finger et al. (2015)
E. moseleyi	12	–	–	–	–	2.10±0.36	–	–	–	Amsterdam is.[q]	2006	Carravieri et al. (2016)
	12	–	–	–	–	1.82±0.30	–	–	–	Amsterdam is.[q]	2007	Carravieri et al. (2016)
	15[f]	–	–	–	–	0.34±0.07	–	–	–	Amsterdam is.[q]	2007	Carravieri et al. (2016)

n/i not informed
[a]Sample collection
[b]Victoria Land (East Antarctica)
[c]South Shetland Islands (West Antarctica)
[d]Subantarctic area of the Atlantic Ocean
[e]Several locations of the Antarctic Peninsula (West Antarctica)
[f]Juvenile
[g]East Antarctica
[h]Authors only report min and max values
[i]Duplicates were performed from the sample
[j](Max 8.16 μg/g)
[k]South East of Indian Ocean
[l]Chicks
[m]Coast of Argentina
[n]Young adults
[o]Coast of Brazil
[p]Victoria, Australia
[q]Southern of the Indian Ocean
[r]Mirror Peninsula (East Antarctica)

Table 2 Mean trace metal levels (μg/g, dw) measured in eggshells of different penguin species

Species	*N*	Al	As	Cd	Pb	Hg	Cu	Zn	Mn	Locations	Date[a]	References
P. adeliae	13	–	–	–	–	0.26 ± 0.08	–	–	–	Terra Nova Bay[b]	1989–1990	Bargagli et al. (1998)
	89	–	–	–	–	0.02 ± 0.01	–	–	–	King George is.[c]	2006–2011	Brasso et al. (2014)
	1	–	–	–	–	0.005	–	8.30	–	Admiralty Bay[c]	2004	Santos et al. (2006)
P. antarctica	92	–	–	–	–	0.07 ± 0.05	–	–	–	King George is.[c]	2006–2011	Brasso et al. (2014)
P. papua	n/i	–	–	–	0.75	0.05	–	–	–	Fildes Peninsula[c]	2002	Yin et al. (2008)
	12	28.96 ± 4.3	<0.3	<0.05	0.68 ± 0.3	–	1.24 ± 0.4	4.07 ± 0.6	0.82 ± 0.08	Livingston is.[c]	2006–2007	Metcheva et al. (2011)
	90	–	–	–	–	0.02 ± 0.01	–	–	–	King George is.[c]	2006	Brasso et al. (2014)

n/i not informed
[a]Sample collection
[b]Victoria Land (East Antarctica)
[c]South Shetland Islands (West Antarctica)

Table 3 Mean trace metal levels (μg/g, dw) in bones of different penguin species

Species	*N*	Al	As	Cd	Pb	Hg	Cu	Zn	Mn	Locations	Date[a]	References
P. adeliae	n/i	–	–	–	1.60	0.02	–	–	–	Zhongshan Station[b]	2001	Yin et al. (2008)
	1	8.49	0.07	0.03	<0.001	–	0.06	138.38	7.44	King George is.[c]	2007–2010	Jerez et al. (2013a)
	1	5.61	0.12	0.17	0.10	–	0.17	106.15	7.56	Avian is.[d]	2007–2010	Jerez et al. (2013a)
	5	11.89 ± 3.69	0.13 ± 0.08	0.01 ± 0.004	0.04 ± 0.1	–	0.96 ± 0.53	227.01 ± 121.11	8.31 ± 3.11	King George is.[c]	2008–2009	Jerez et al. (2013b)
P. antarctica	2	4.16 ± 1.02	0.08 ± 0.07	<0.001	<0.001	–	0.17 ± 0.22	221.3 ± 9.19	6.66 ± 0.57	King George is.[c]	2007–2010	Jerez et al. (2013a)
	4	7.30 ± 6.09	0.04 ± 0.01	0.07 ± 0.03	0.21 ± 0.12	–	0.19 ± 0.1	138.77 ± 20.63	8.40 ± 1.25	Deception is. [c]	2007–2010	Jerez et al. (2013a)
	5	7.38 ± 2.93	0.14 ± 0.13	0.004 ± 0.001	0.14 ± 0.02	–	0.71 ± 0.36	235.01 ± 40.62	12.5 ± 2.13	Deception is. [c]	2008–2009	Jerez et al. (2013b)
P. papua	1	69.95	0.13	0.001	0.19	–	0.79	184.1	11.01	King George is.[c]	2008–2009	Jerez et al. (2013b)
	13	32.28 ± 14.96	0.19 ± 0.04	0.008 ± 0.003	<0.001	–	1.15 ± 0.33	244.6 ± 7.99	18.35 ± 2.28	Byers peninsula[c]	2009	Barbosa et al. (2013)
	8	18.63 ± 4.44	0.12 ± 0.01	0.005 ± 0.001	0.02 ± 0.01	–	0.74 ± 0.09	223.82 ± 21.91	9.81 ± 1.21	Hannah point[c]	2009	Barbosa et al. (2013)
	2	7.59 ± 0.2	0.06 ± 0.01	0.002 ± 0.002	0.15 ± 0.19	–	0.20 ± 0.11	180.05 ± 29.4	8.27 ± 1.78	King George is.[c]	2007–2010	Jerez et al. (2013a)
	1[e]	–	–	0.10 ± 0.02	0.30 ± 0.14	–	0.90 ± 0.028	81 ± 0.6	2.50 ± 0.23	Livingston is.[c]	2002–2006	Metcheva et al. (2010)

[a]Sample collection
[b]Mirror Peninsula (East Antarctica)
[c]South Shetland Islands (West Antarctica)
[d]Southern of the Antarctic Peninsula
[e]Duplicates were performed from the sample

Table 4 Mean trace metal levels (μg/g, dw) in kidneys of different penguin species

Species	N	Al	As	Cd	Pb	Hg	Cu	Zn	Mn	Locations	Date[a]	References
P. adeliae	1	–	–	–	–	1.20	–	–	–	Terra Nova Bay[b]	1989–1990	Bargagli et al. (1998)
	1	0.74	1.07	54.41	<0.01	–	10.74	163.71	3.78	King George is.[c]	2007–2010	Jerez et al. (2013a)
	1[d]	3.48	0.45	0.68	<0.01	–	12.66	119.90	7.79	King George is.[c]	2007–2010	Jerez et al. (2013a)
	2	14.12 ± 3.86	0.38 ± 0.12	351.8 ± 0.08	0.21 ± 0.17	–	14.78 ± 3.04	234.3 ± 62.24	5.77 ± 1.36	Avian is.[e]	2007–2010	Jerez et al. (2013a)
	5	4.09 ± 7.05	0.44 ± 0.24	0.20 ± 0.15	0.05 ± 0.12	–	11.85 ± 3.69	85.74 ± 19.49	11.18 ± 6.12	King George is.[c]	2008–2009	Jerez et al. (2013b)
	3	–	0.547 ± 0.033	0.339 ± 0.012	0.144 ± 0.007	0.146 ± 0.004	1.6 ± 0.12	–	9.4 ± 0.2	Potter Cove[f]	2002–2003	Smichowski et al. (2006)
	5	–	–	263.8 ± 216.6	–	–	17.80 ± 4.1	–	–	Weddell Sea[g]	1982–1983	Steinhagen-Schneider (1986)
P. antarctica	2	0.69 ± 0.38	0.52 ± 0.6	0.49 ± 0.32	<0.01	–	17.13 ± 2.63	107.79 ± 23.57	10.13 ± 2.37	King George is.[c]	2007–2010	Jerez et al. (2013a)
	4	0.75 ± 0.76	0.58 ± 0.12	263.93 ± 139.77	0.18 ± 0.01	–	15.33 ± 4.67	149.8 ± 49.23	5.35 ± 0.75	Deception is.[c]	2007–2010	Jerez et al. (2013a)
	5	10.93 ± 10.57	0.50 ± 0.09	0.54 ± 0.29	0.14 ± 0.02	–	13.64 ± 2.28	92.83 ± 32.19	10.19 ± 2.63	Deception is.[c]	2008–2009	Jerez et al. (2013b)
A. forsteri	4	–	–	270.2 ± 126.8	–	–	19.1 ± 3.0	–	–	Weddell Sea[g]	1982–1983	Steinhagen-Schneider (1986)
P. papua	3	2.13 ± 0.62	0.67 ± 0.41	11.37 ± 14.1	0.07 ± 0.03	–	13.99 ± 2.91	93.14 ± 42.13	6.40 ± 3.07	King George is.[c]	2007–2010	Jerez et al. (2013a)
	4[d]	4.8 ± 4.24	0.43 ± 0.17	1.54 ± 0.71	<0.01	–	19.99 ± 6.83	152.14 ± 18.51	7.33 ± 3.38	King George is.[c]	2007–2010	Jerez et al. (2013a)
	5	6.91 ± 3.95	0.40 ± 0.23	0.20 ± 0.05	<0.001	–	14.26 ± 4.33	125.43 ± 12.60	7.54 ± 3.47	King George is.[c]	2008–2009	Jerez et al. (2013b)
	1[h]	–	–	41.20 ± 0.67	0.10 ± 0.5	–	8.10 ± 0.45	232.0 ± 2.67	4.90 ± 0.44	Livingston is.[c]	2002–2006	Metcheva et al. (2010)
S. magellanicus	22	–	–	46.50 ± 33.55	0.55 ± 0.3	2.47 ± 1.42	–	–	–	Rio Grande do Sul[i]	n/i	Kehrig et al. (2015)

[a]Sample collection
[b]Victoria Land (East Antarctica)
[c]South Shetland Islands (West Antarctica)
[d]Juvenile
[e]Several locations of the Antarctic Peninsula
[f]King George Island (South Shetland Islands, West Antarctica)
[g]Northeast of the Antarctic Peninsula
[h]Duplicates were performed from the sample
[i]Coast of Brazil

Table 5 Mean trace metal levels (μg/g, dw) in liver of different penguin species

Species	*N*	Al	As	Cd	Pb	Hg	Cu	Zn	Mn	Locations	Date[a]	References
P. adeliae	1	4.19	1.20	4.41	0.05	–	10.91	136.3	8.58	King George is.[b]	2007–2010	Jerez et al. (2013a)
	2	0.55 ± 0.11	0.33 ± 0.04	22.03 ± 10.47	0.06 ± 0.03	–	15.34 ± 1.87	141.75 ± 4.21	11.55 ± 4.55	Avian is.[c]	2007–2010	Jerez et al. (2013a)
	1[d]	1.93	0.30	0.18	<0.01	–	22.89	182.58	15.83	King George is.[b]	2007–2010	Jerez et al. (2013a)
	5	6.81 ± 11.91	0.60 ± 0.4	0.06 ± 0.05	0.04 ± 0.07	–	92.06 ± 74.53	133.88 ± 71.42	12.01 ± 5.8	King George is.[b]	2008–2009	Jerez et al. (2013b)
	3	–	0.499 ± 0.024	0.102 ± 0.007	0.202 ± 0.009	0.269 ± 0.01	18.0 ± 1.0	–	10.0 ± 0.2	Potter Cove[e]	2002–2003	Smichowski et al. (2006)
	3	–	–	7.20 ± 0.12	0.30 ± 0.05	–	11.90 ± 0.5	140.0 ± 4	6.80 ± 0.2	AP[f]	1989	Szefer et al. (1993)
	5	–	–	7.50 ± 2.4	–	–	19.90 ± 5.8	–	–	Weddell Sea[g]	1982–1983	Steinhagen-Schneider (1986)
P. antarctica	2	1.0 ± 0.14	0.37 ± 0.36	0.16 ± 0.08	0.05 ± 0.01	–	24.26 ± 11.18	330.34 ± 293.26	14.76 ± 4.17	King George is.[b]	2007–2010	Jerez et al. (2013a)
	4	2.02 ± 1.47	0.67 ± 0.15	27.54 ± 14.47	0.15 ± 0.06	–	14.95 ± 0.67	126.05 ± 25.18	9.30 ± 2.06	Deception is.[b]	2007–2010	Jerez et al. (2013a)
	5	15.52 ± 15.55	0.47 ± 0.14	0.11 ± 0.08	0.18 ± 0.02	–	95.10 ± 48.67	132.2 ± 64.4	11.42 ± 3.24	Deception is.[b]	2008–2009	Jerez et al. (2013b)
	3	–	–	10.70 ± 0.3	0.01	–	12.60 ± 0.4	126.0 ± 1	7.50 ± 0.5	AP[f]	1989	Szefer et al. (1993)
A. forsteri	4	–	–	27.7 ± 15.6	–	–	23.4 ± 4.0	–	–	Weddell Sea[g]	1982–1983	Steinhagen-Schneider (1986)
P. papua	3	2.19 ± 0.52	1.01 ± 0.9	1.05 ± 1.43	0.10 ± 0.07	–	102.57 ± 155.93	112.56 ± 72.69	7.71 ± 6.34	King George is.[b]	2007–2010	Jerez et al. (2013a)
	4[d]	1.62 ± 1	0.79 ± 0.63	0.40 ± 0.18	<0.01	–	386.13 ± 174.48	237.19 ± 22.38	8.30 ± 0.57	King George is.[b]	2007–2010	Jerez et al. (2013a)
	5	2.12 ± 2.05	0.45 ± 0.18	0.08 ± 0.04	<0.001	–	142.4 ± 63.85	152.91 ± 45.53	10.51 ± 3.74	King George is.[b]	2008–2009	Jerez et al. (2013b)
	3	–	–	3.19 ± 0.1	0.48 ± 0.12	–	26.50 ± 0.9	100.0 ± 4	8.80 ± 0.8	AP[f]	1989	Szefer et al. (1993)

	1[h]	–	–	2.32 ± 0.27	0.50	–	24.70 ± 0.19	72 ± 0.53	7.10 ± 0.17	Livingston is.[b]	2002–2006	Metcheva et al. (2010)
S. magellanicus	22	–	–	7.25 ± 4.71	0.58 ± 0.32	5.70 ± 3.73	–	–	–	Rio Grande do Sul[i]	n/i	Kehrig et al. (2015)

[a]Sample collection
[b]South Shetland Islands (West Antarctica)
[c]Antarctic Peninsula
[d]Juvenile
[e]King George Island (South Shetland Islands, West Antarctica)
[f]Several locations of the Antarctic Peninsula
[g]Northeast of the Antarctic Peninsula
[h]Duplicates were performed from the sample
[i]Coast of Brazil

Table 6 Mean trace metal levels (μg/g, dw) in muscles of different penguin species

Species	*N*	Al	As	Cd	Pb	Hg	Cu	Zn	Mn	Locations	Date[a]	References
P. adeliae	1	–	–	–	–	0.6	–	–	–	Terra Nova Bay[b]	1989	Bargagli et al. (1998)
	1	3.27	0.37	1.09	<0.01	–	7.43	149.95	0.63	King George is.[c]	2007–2010	Jerez et al. (2013a)
	2	1.78 ± 1.91	0.30 ± 0.16	2.63 ± 2.09	0.15 ± 0.11	–	8.53 ± 2.41	66.26 ± 57.77	1.11 ± 0.39	Avian is.[d]	2007–2010	Jerez et al. (2013a)
	1[e]	3.41	0.18	<0.01	<0.01	–	6.97	163.75	0.91	King George is.[b]	2007–2010	Jerez et al. (2013a)
	5	6.14 ± 6.72	0.39 ± 0.25	0.01 ± 0.02	0.04 ± 0.01	–	5.52 ± 1.97	104.34 ± 49.7	1.13 ± 0.4	King George is.[b]	2008–2009	Jerez et al. (2013b)
	3	–	0.815 ± 0.007	<0.001	0.12 ± 0.007	–	6.4 ± 0.4	–	1.50 ± 0.1	Potter Cove[f]	2002–2003	Smichowski et al. (2006)
	9	–	–	0.46 ± 0.42	0.04 ± 0.01	–	7.90 ± 0.7	46.7 ± 10	0.51 ± 0.16	AP[g]	1989	Szefer et al. (1993)
	5	–	–	0.32 ± 0.02	–	–	8.20 ± 2.4	–	–	Weddell Sea[h]	1982	Steinhagen-Schneider (1986)
P. antarctica	2	1.07 ± 0.36	0.57 ± 0.62	0.01 ± 0.001	<0.01	–	8.07 ± 1.28	139.91 ± 40.94	0.87 ± 0.28	King George is.[c]	2007–2010	Jerez et al. (2013a)
	4	12.32 ± 10.04	1.04 ± 0.27	1.83 ± 0.63	0.17 ± 0.08	–	6.69 ± 1.73	118.8 ± 40.73	1.17 ± 0.68	Deception is.[c]	2007–2010	Jerez et al. (2013a)
	5	114.88 ± 125.59	0.59 ± 0.3	0.01 ± 0.01	0.20 ± 0.06	–	6.82 ± 1.2	105.08 ± 55.41	2.55 ± 1.53	Deception is.[c]	2008–2009	Jerez et al. (2013b)
	3	–	–	0.57 ± 0.02	0.01	–	9.70 ± 0.3	37.0 ± 2.9	0.76 ± 0.31	AP[g]	1989	Szefer et al. (1993)
P. papua	3	1.39 ± 0.95	0.63 ± 0.53	0.11 ± 0.18	0.18 ± 0.05	–	7.97 ± 1.15	103.07 ± 60.55	0.85 ± 0.39	King George is.[c]	2007–2010	Jerez et al. (2013a)
	4[e]	2.01 ± 1.96	0.36 ± 0.21	0.01 ± 0.01	<0.01	–	9.95 ± 2.08	139.39 ± 46.68	0.52 ± 0.06	King George is.[c]	2007–2010	Jerez et al. (2013a)
	5	43.71 ± 21.93	0.40 ± 0.23	0.01 ± 0.01	<0.001	–	4.43 ± 1.46	106.6 ± 37.42	1.46 ± 0.43	King George is.[c]	2008–2009	Jerez et al. (2013b)
	3	–	–	0.02	0.01	–	8.20 ± 0.5	35.7 ± 3	0.46 ± 0.03	AP[g]	1989	Szefer et al. (1993)
	1[i]	–	–	0.5 ± 0.12	<0.60	–	5.60 ± 0.31	24 ± 0.24	1.4 ± 0.13	Livingston is.[c]	2002–2006	Metcheva et al. (2010)

A. forsteri	4	–	–	0.35 ± 0.09	–	–	5.50 ± 2	–	–	Weddell Sea[h]	1982–1983	Steinhagen-Schneider (1986)

n/*i* not informed
[a]Sample collection
[e]Juvenile
[i]Duplicates were performed from the sample
[b]Victoria Land (East Antarctica)
[c]South Shetland Islands (West Antarctica)
[d]Antarctic Peninsula
[f]King George Island (South Shetland Islands, West Antarctica)
[g]Antarctic Peninsula (locations not specified)
[h]Northeast of the Antarctic Peninsula

Table 7 Mean trace metal levels (μg/g, dw) in stomach contents of different penguin species

Species	*N*	Al	As	Cd	Pb	Hg	Cu	Zn	Mn	Locations	Date[a]	References
P. adeliae	5	–	–	–	–	0.08 ± 0.01	–	–	–	Terra Nova Bay[b]	1989–1990	Bargagli et al. (1998)
	1	349.72	0.47	0.45	0.07	–	4.85	26.57	6.64	King George is.[c]	2007–2010	Jerez et al. (2013a)
	2	46.80 ± 54.31	3.22 ± 0.06	1.10 ± 0.8	0.28 ± 0.19	–	66.42 ± 34.43	38.99 ± 14.05	2.20 ± 0.11	Avian is.[d]	2007–2010	Jerez et al. (2013a)
	5	282.01 ± 235.63	2.00 ± 1.58	0.23 ± 0.17	0.40 ± 0.26	–	57.81 ± 35.82	71.16 ± 48.82	10.57 ± 8.76	King George is.[c]	2008–2009	Jerez et al. (2013b)
	45	–	–	2.90	0.20	0.10	–	–	–	Edmonson Point[b]	1995	Ancora et al. (2002)
P. antarctica	2	641.07 ± 255.18	1.44 ± 1.83	0.17 ± 0.06	0.03 ± 0.04	–	51.07 ± 49.14	49.39 ± 8.3	9.33 ± 4.97	King George is.[c]	2007–2010	Jerez et al. (2013a)
	1	193.52	1.77	0.71	0.12	–	54.86	46.67	5.99	Deception is.[c]	2007–2010	Jerez et al. (2013a)
	5	477.85 ± 192.75	1.92 ± 1.11	0.32 ± 0.34	0.33 ± 0.11	–	65.67 ± 50.01	31.04 ± 10.02	12.4 ± 6.46	Deception is.[c]	2008–2009	Jerez et al. (2013b)
P. papua	2	2594.6 ± 1306.7	2.0 ± 0.09	0.09 ± 0.11	0.71 ± 0.42	–	30.51 ± 35.73	19.84 ± 4.63	82.43 ± 27.49	King George is.[c]	2007–2010	Jerez et al. (2013a)
	4[e]	854.88 ± 1000.14	0.28 ± 0.07	0.12 ± 0.07	0.05 ± 0.02	–	7.33 ± 1.13	41.09 ± 16.40	16.27 ± 15.69	King George is.[c]	2007–2010	Jerez et al. (2013a)
	5	2010.1 ± 3231.8	2.04 ± 2.92	0.24 ± 0.15	0.17 ± 0.14	–	58.69 ± 28.48	31.46 ± 12.52	36.89 ± 66.39	King George is.[c]	2008–2009	Jerez et al. (2013b)

[a]Sample collection
[b]Victoria Land (East Antarctica)
[c]South Shetland Islands (West Antarctica)
[d]Southern of the Antarctic Peninsula
[e]Juvenile

Table 8 Mean trace metal levels (μg/g, dw) in excreta of different penguin species

Species	*N*	Al	As	Cd	Pb	Hg	Cu	Zn	Mn	Locations	Date[a]	References
P. adeliae	7	–	–	–	–	0.17 ± 0.1	–	–	–	Terra Nova Bay[b]	1989–1990	Bargagli et al. (1998)
	14	–	–	5.5	0.30	0.20	–	–	–	Edmonson Point[b]	1995	Ancora et al. (2002)
	n/i	–	–	–	0.5–3.7	0.15–0.25	–	–	–	Zhongshan Station[c]	2001	Yin et al. (2008)
	27	–	1.14 ± 0.39	3.96 ± 2.36	1.96 ± 0.86	0.52 ± 0.31	558.9 ± 217.73	262.7 ± 91.89	–	Arctowski[d]	2012–2013	Celis et al. (2015a)
	18	–	0.95 ± 0.41	2.77 ± 0.92	1.53 ± 0.75	0.40 ± 0.18	585.8 ± 196.22	215.8 ± 42.47	–	Kopaitic Island[e]	2012–2013	Celis et al. (2015a)
	10	–	0.72 ± 0.47	1.78 ± 0.39	0.59 ± 0.5	0.10 ± 0.08	402.9 ± 54.76	215.7 ± 91.18	–	Yalour is.[e]	2012–2013	Celis et al. (2015a)
	10	–	0.66 ± 0.53	1.63 ± 0.43	0.45 ± 0.39	0.13 ± 0.08	362.9 ± 38	188.4 ± 41.82	–	Avian is.[e]	2012–2013	Celis et al. (2015a)
P. antarctica	3	–	–	0.16 ± 0.01	3.80 ± 0.2	–	37.6 ± 2.0	456.0 ± 4.0	138 ± 8	AP[f]	1989	Szefer et al. (1993)
	n/i	–	–	–	1.0–1.8	0.06–0.15	–	–	–	Barton Peninsula[d]	2000	Yin et al. (2008)
	4	–	0.43 ± 0.24	3.30 ± 0.18	1.06 ± 0.6	–	168.9 ± 40.8	295.7 ± 59.01	–	Hydrurga Rocks[e]	2011–2012	Espejo et al. (2014)
	10	–	0.40 ± 0.15	1.89 ± 0.35	1.07 ± 1.5	–	229.9 ± 39.23	246.8 ± 53.37	–	Cape Shirreff[d]	2011–2012	Espejo et al. (2014)
	9	–	0.70 ± 0.26	3.13 ± 0.59	1.31 ± 0.78	–	259.99 ± 79.51	227.8 ± 63.9	–	Narebski Point[d]	2011–2012	Espejo et al. (2014)
	9	–	0.55 ± 0.31	1.88 ± 0.65	1.27 ± 0.35	–	286.7 ± 85.75	210.0 ± 115.58	–	Kopaitic Island[e]	2011–2012	Espejo et al. (2014)
P. papua	n/i	–	–	–	0.11	0.15	–	–	–	Fildes Peninsula[d]	2002	Yin et al. (2008)
	10	316 ± 47.5	5.13 ± 1.79	1.03 ± 0.36	<0.4	–	104.0 ± 2.1	145.0 ± 2.9	12.3 ± 1.2	Livingston is.[d]	2006–2007	Metcheva et al. (2011)
	10	–	0.33 ± 0.22	2.51 ± 0.89	2.89 ± 1.07	–	199.95 ± 62.47	379.99 ± 82.73	–	Base O'Higgins[e]	2011–2012	Espejo et al. (2014)
	4	–	0.44 ± 0.38	2.15 ± 0.47	0.78 ± 0.22	–	114.7 ± 41.55	192.2 ± 39.32	–	Yankee Harbor[d]	2011–2012	Espejo et al. (2014)

(continued)

Table 8 (continued)

Species	*N*	Al	As	Cd	Pb	Hg	Cu	Zn	Mn	Locations	Date[a]	References
	3	–	0.37 ± 0.28	3.35 ± 0.23	2.55 ± 1.02	–	184.5 ± 10.81	324.3 ± 106.51	–	Mikkelsen Harbor[e]	2011–2012	Espejo et al. (2014)
	4	–	0.43 ± 0.13	2.16 ± 0.35	0.87 ± 0.53	–	130.05 ± 20.02	195.38 ± 32.84	–	Danco is.[e]	2011–2012	Espejo et al. (2014)
	10	–	0.36 ± 0.29	2.14 ± 0.91	2.74 ± 1.3	–	222.51 ± 85.48	201.2 ± 63.39	–	Base G. Videla[e]	2011–2012	Espejo et al. (2014)
	5	–	0.52 ± 0.16	2.40 ± 0.83	2.05 ± 1.95	–	154.2 ± 28.41	172.92 ± 62.85	–	Yelcho Station[e]	2011–2012	Espejo et al. (2014)
	4	–	0.33 ± 0.17	1.98 ± 0.11	2.51 ± 0.84	–	148.8 ± 32.3	246.95 ± 30.37	–	Brown Station[e]	2011–2012	Espejo et al. (2014)
	10	–	0.15 ± 0.097	0.73 ± 0.27	0.74 ± 0.953	6.60 ± 4.153	–	–	–	O'Higgins Base[e]	2011	Celis et al. (2012)
	10	–	0.38 ± 0.176	1.72 ± 0.832	0.34 ± 0.388	1.15 ± 0.828	–	–	–	Base G. Videla[e]	2011	Celis et al. (2012)
	11	–	–	1.58 ± 1.11	0.08 ± 0.08	–	146.0 ± 76.17	142.97 ± 35.51	22.43 ± 8.57	Neko Harbor[b]	2014	Celis et al. (2015b)
	10	–	–	1.24 ± 0.25	0.09 ± 0.1	–	201.5 ± 64.14	108.74 ± 25.23	17.84 ± 13.22	Doumer is.[e]	2014	Celis et al. (2015b)
	10	–	–	1.97 ± 0.86	1.46 ± 0.49	–	222.51 ± 85.48	201.2 ± 63.39	36.62 ± 16.97	Stranger Point[d]	2014	Celis et al. (2015b)
	10	–	–	2.92 ± 0.81	1.68 ± 0.58	–	266.83 ± 42.77	317.92 ± 46.6	44.75 ± 10.67	Base O'Higgins[e]	2014	Celis et al. (2015b)
S. humboldti	20	–	1.84 ± 2.65	47.7 ± 38.71	1.80 ± 0.3	0.77 ± 0.83	147.79 ± 146.42	487.11 ± 395.15	–	Pan de Azúcar is.[g]	2011–2012	Celis et al. (2014)
	19	–	0.36 ± 0.4	21.24 ± 18.35	1.59 ± 2.12	0.46 ± 0.19	69.62 ± 24.98	222.55 ± 59.2	–	Chañaral Island[g]	2011–2012	Celis et al. (2014)
	24	–	7.86 ± 4.88	42.47 ± 45.55	12.79 ± 9.97	0.61 ± 0.4	199.67 ± 81.78	0.83 ± 0.33	–	Cachagua Island[g]	2011–2012	Celis et al. (2014)
A. forsteri	5	–	–	–	–	0.31 ± 0.03	–	–	–	Terra Nova Bay[b]	1989–1990	Bargagli et al. (1998)
A. patagonicus	n/i	–	–	–	0.6–1.1	0.25–0.35	–	–	–	Zhongshan Station[c]	2001	Yin et al. (2008)

n/i not informed
[a]Sample collection
[b]Victoria Land (East Antarctica)
[c]Mirror Peninsula (East Antarctica)
[d]South Shetland Islands (West Antarctica)
[e]Antarctic Peninsula (West Antarctica)
[f]Antarctic Peninsula (locations not specified)
[g]Coast of Chile

Table 9 Mean trace metal levels (μg/g, dry weight) in blood, brain, testicles, embryo, spleen, and heart of different penguin species

Matrix	*N*	Species	Al	As	Cd	Pb	Hg	Cu	Zn	Mn	Locations	Date[a]	References
Blood	10	*E. minor*	3.89 ± 1.26	3.72 ± 1.76	–	0.07 ± 0.02	2.75 ± 0.85	2.48 ± 0.44	37.97 ± 5.28	–	St. Kilda[b]	2012	Finger et al. (2015)
Blood	11	*E. minor*	3.19 ± 0.84	1.07 ± 1.22	–	0.04 ± 0.01	0.86 ± 0.23	2.14 ± 0.42	33.47 ± 3.27	–	Phillip is.[b]	2012	Finger et al. (2015)
Blood	10	*E. minor*	4.22 ± 1.67	0.67 ± 0.43	–	0.04 ± 0.01	0.84 ± 0.37	2.32 ± 0.40	38.77 ± 6.76	–	Notch is.[b]	2012	Finger et al. (2015)
Brain	1	*P. adeliae*	–	–	–	–	0.43	–	–	–	Terra Nova Bay[c]	1989–1990	Bargagli et al. (1998)
Testicles	1	*P. adeliae*	–	–	–	–	0.42	–	–	–	Terra Nova Bay[c]	1989–1990	Bargagli et al. (1998)
Embryo	12	*P. papua*	14.56 ± 2.4	<0.3	<0.05	<0.4	–	2.82 ± 0.7	25.27 ± 2.5	0.67 ± 0.06	Livingston is.[d]	2006–2007	Metcheva et al. (2011)
Spleen	1[e]	*P. papua*	–	–	3.5 ± 0.41	< 0.95	–	24.7 ± 0.19	232.0 ± 2.67	6.30 ± 0.15	Livingston is.[d]	2002–2006	Metcheva et al. (2010)
Heart	1[e]	*P. papua*	–	–	0.1 ± 0.02	0.20 ± 0.09	–	11.3 ± 0.09	91.5 ± 0.67	1.0 ± 0.25	Livingston is.[d]	2002–2006	Metcheva et al. (2010)

[a]Sample collection
[b]Victoria (Australia)
[c]Victoria Land (East Antarctica)
[d]South Shetland Islands (West Antarctica)
[e]Duplicates were performed from the sample

feathers of pygoscelid penguins from Antarctica. Metals such as Mo, V, or Y have not been reported in penguins.

Feathers constitute the most common biological matrix used in situ for determining trace elements in penguins (Table 1). Metals are delivered mainly by the blood supply, which is linked to the feeding habits of the bird (Metcheva et al. 2006). Some evidence shows that the concentration of Hg in feathers reflects levels in the blood during formation (Dauwe et al. 2005). Trace element burdens in feathers express past exposure and accumulation during the inter-moult period, thus they are more representative of long-term rather than acute exposure, at least for Hg (Furness et al. 1986). Reports in penguin feathers comprise ten species, most of which are from Antarctica and subantarctic islands (Table 1). Similarly, there is plenty of information of trace elements in penguin guano, particularly from Antarctica (Table 8), but there are few data on other species that live in lower latitudes, except for a study of Humboldt penguins (*Spheniscus humboldti*) from the coast of Chile (Celis et al. 2014).

There is very little information on trace elements in penguin eggshells (Table 2), bones (Table 3), and kidneys (Table 4). Concentrations of trace elements in blood, brain, testicles, embryo, spleen, and heart of penguins have been poorly investigated (Table 9). Studies of trace elements in the liver of penguins (Table 5) correspond mostly to species that inhabit the Antarctica and subantarctic islands. Data of trace elements in muscles of penguins are scarce and they are exclusively focused on species that inhabit Antarctica (Table 6). Studies on metals in stomach contents of penguins are scarce and all of them have been carried out in Antarctica (Table 7).

Trace elements are chemicals that occur in natural and perturbed environments in small amounts (Prashanth et al. 2016). Their inadequate intake can damage the function of cells, causing physiological disorders and disease (Soria et al. 1995). These chemicals can be classified according to their biological significance, as non-essential and essential trace elements. Non-essential elements, such as Pb, Be, Cd, Hg, As, Sb, and Ti, have no known function in the animal body, and their presence may produce toxicity. Essential elements (Cr, Co, Cr, Cu, Fe, Mo, Se, Zn, and Mn) are required in small amounts because they perform vital functions for the maintenance of animal life, growth, and reproduction (Nordberg and Nordberg 2016). Some trace elements such as Ni, Sn, V, and Al cannot be yet classified as essential, as their role is not clear in animals, including humans (Prashanth et al. 2016). In general, the information available on concentrations of trace elements is fragmented in time and space, so it is not possible to build trends. Therefore, implementations of monitoring programs that incorporate these variables are required.

3.1 Non-essential Trace Elements

3.1.1 Aluminum

The maximum concentrations of Al have been found in stomach contents of gentoo penguins from King George Island, Antarctica (2595 μg/g, Table 7) and in feathers of adult chinstrap penguins from Deception Island, Antarctica (203.13 μg/g, Table 1). In contrast, the lowest concentration of Al (0.55 μg/g) has been reported in livers of Adélie penguins from Avian Island, Antarctica (Table 5). The high Al levels found in penguins from King George and Deception Islands could be linked to the abundance of this metal in bioavailable forms in the sediments of these areas (dos Santos et al. 2005; Deheyn et al. 2005).

Concentrations of Al in penguin feathers (0.71–203.13 μg/g) are highest in adult chinstrap penguins at Deception Island, whereas the lowest concentrations are in juvenile Adélie penguins at Avian Island (Table 1). This range is lower than the concentrations of Al (96–866 μg/g) in feathers of birds from Europe and North America (Rattner et al. 2008; Lucia et al. 2010). Highest concentrations of Al (866 μg/g) have been measured in feathers of osprey eagles (*Pandion haliaetus*) (Rattner et al. 2008).

Only one study reports the concentration of Al in penguin eggshells (28.96 μg/g, Table 2), which is higher than that found by Custer et al. (2007) in seagulls of North America (3.3 μg/g).

Concentrations of Al in penguin bones (4.16–69.95 μg/g) are the highest in gentoo penguins from King George Island, while the lowest concentrations are in chinstrap penguins from the same location (Table 3). Studies on levels of Al in bones of seabirds are scarce, and the few data in penguins are all from species from genera *Pygoscelis* and *Aptenodytes*, being higher than those of birds from the Northern Hemisphere (1.37–6.9 μg/g, Dauwe et al. 2005).

In kidneys, Al concentrations (0.69–14.12 μg/g, Table 4) are highest in Adélie penguins from Avian Island and are lowest in chinstrap penguins from King George Island. In comparison, Al levels in kidneys of aquatic birds from the Southwest coast of France (6.1–8.9 μg/g, Lucia et al. 2010) are within the range reported in the penguin kidneys from Antarctica.

Concentrations of Al in the liver (0.55–15.52 μg/g, Table 5) are highest in chinstrap penguins from Deception Island (Table 5), whereas the lowest concentrations of Al correspond to Adélie penguins from Avian Island. There is little information about the levels of Al in livers of seabirds, although it is possible to see that the concentrations of Al in penguins are within the range reported in the aquatic and terrestrial birds from Europe (0.18–37.3 μg/g) (Scheuhammer 1987; Dauwe et al. 2005).

In muscles, Al concentrations (1.07–114.88 μg/g, Table 6) are highest in chinstrap penguins from Deception Island, whereas the lowest concentrations are in the same species from King George Island. Despite the lack of data on seabirds from other regions, levels of Al in penguin muscles are higher than those found in a study carried out from Europe in muscles of the Great tit (0.08–1.46 μg/g, Dauwe et al. 2005).

Concentrations of Al in penguin stomachs (46.80–2594.6 μg/g, Table 7) are highest in gentoo penguins from King George Island, whereas the lowest concentrations are in Adélie penguins from Avian Island. This range is higher than the concentrations of Al (0.22–23.5 μg/g) in stomach contents of wild birds from Europe (Dauwe et al. 2005).

There is only one measurement of Al in excreta (316 μg/g, Table 8), which is from gentoo penguins at Livingston Island, showing a deficit of information for this element. No Al was found in guano of other seabirds. Birds are most likely exposed to Al through their diets, and most Al is excreted via the feces and only a fraction is retained (Sparling and Lowe 1996).

In blood, the only existing study corresponds to the little penguin (*Eudyptula minor*), and Al concentrations (3.19–4.22 μg/g, Table 9) present less variability and are within the range reported in the seagulls of the Northern Hemisphere (1.34–4.11 μg/g, Kim et al. 2013).

The main toxic effects of Al that have been reported in animals are produced in the central nervous system, though long-lasting exposures can also affect the skeletal system, decreasing its rate of formation and increasing the risk of fractures. The functioning of the renal, endocrine, reproductive, and cardiac systems is also affected by a chronic exposure to this metal (Sjögren et al. 2007). In birds, Al is poorly absorbed and its potential for toxicity is low, thus Al levels in soft tissue do not necessarily reflect toxicity to the individual (Scheuhammer 1987). Nevertheless, there is some evidence indicating that Al found in the bone marrow tissue of humeri of wild pied flycatchers (*Ficedula hypoleuca*) can produce small clutches, defective eggshell formation, and intrauterine bleeding, similar to the symptoms of Al intoxication in mammals (Nyholm 1981). Al interferes with the deposition of Ca, resulting in weak bones and eggs, besides affecting the reproductive capacity (Nayak 2002). No studies have been performed in penguins to determine any possible effects produced by this metal.

3.1.2 Arsenic

The maximum As concentrations in tissues and organs are in the liver of Adélie penguins (*Pygoscelis adeliae*) from King George Island (1.2 μg/g, Table 5) and kidneys of the same species and location (1.07 μg/g, Table 4). In contrast, the lowest concentration of As (0.01 μg/g) has been reported in feathers of adult chinstrap penguins (*Pygoscelis antarctica*) from Livingston Island, Antarctica (Table 1). Arsenic tends to accumulate in almost all organs, mainly in the liver where biomethylation of As takes place producing some kind of acids, such as monomethylarsonic and dimethylarsonic (Khan et al. 2014).

Concentrations of As in feathers (0.01–0.88 μg/g, Table 1) are highest in adult gentoo penguins that inhabit Livingston Island, and are lowest in adult chinstrap penguins from Livingston Island. Arsenic concentrations in feathers of black-legged kittiwakes (*Rissa tridactyla*) and black oystercatchers (*Haematopus bachmani*) from Alaska (0.17–0.34 μg/g, Burger et al. 2008) and in feathers of black-tailed gull chicks (*Larus crassirostris*) from Korea (0.15–0.44 μg/g, Kim et al. 2013) are within the range found in the penguins. There is little information on the levels of As in penguin eggshells (Table 2).

In bones, concentrations of As are highest in gentoo penguins that inhabit Byers Peninsula (Table 3), while the lowest concentrations are in chinstrap penguins from Deception Island. Concentrations of As in penguin bones (0.04–0.19 μg/g) are within the range reported in the aquatic and terrestrial birds of the Northern Hemisphere (<0.0001–1.60 μg/g, Lebedeva 1997).

In kidneys, As concentrations (0.38–1.07 μg/g, Table 4) are highest in Adélie penguins from King George Island and lowest in the same species from Avian island. Arsenic levels in penguin kidneys are within the range found in the kidneys of passerine birds from the Northern Hemisphere (0.071–1.81 μg/g, Sánchez-Virosta et al. 2015).

In the liver, the content of As (0.30–1.20 μg/g) is highest in adult Adélie penguins from King George Island, and the lowest in juvenile same species, location, and sampling date (Table 5). The highest levels of As in adult penguins is probably due to the fact that this element accumulates in the animal body, and thus its level is directly related to the age of the individuals (Khan et al. 2014). The concentrations of As in penguin livers are lower than those found in seabirds of the Northern Hemisphere (0.22–5.62 μg/g) (Lucia et al. 2010; Ribeiro et al. 2009; Skoric et al. 2012).

In muscles, As concentrations (0.18–1.04 μg/g, Table 6) are highest in chinstrap penguins from Deception Island, and lowest in Adélie penguins from King George Island. The concentrations of As in penguin muscles are higher than those reported in wild birds from the Northern Hemisphere (0.01–0.35 μg/g, Gasparik et al. 2010).

In the stomach, As concentrations are highest in Adélie penguins from Avian Island, and are lowest in the same species from King George Island (Table 7). Arsenic concentrations in penguin stomach contents (0.28–3.22 μg/g) are higher than those levels found in wild birds from Europe (0.006–0.76 μg/g, Dauwe et al. 2005).

In excreta, As concentrations are highest in Humboldt penguins from Cachagua Island (Chile), and lowest in gentoo penguins from O'Higgins Base, Antarctic Peninsula (Table 8). In general, levels of As in penguin guano (0.15–7.86 μg/g) are lower than those levels found in guano of wild birds from the Northern Hemisphere (0.42–16.03 μg/g, Dauwe et al. 2000; Kler et al. 2014).

In blood, the highest As concentration is in the little penguin from Australia (Table 9). Levels of As in penguin blood (0.67–3.72 μg/g) are higher than those levels observed in black-tailed gull chicks of the Northern Hemisphere (0.26–0.48 μg/g, Kim et al. 2013).

Generally, in birds As is initially accumulated in liver and kidneys and subsequently it is redistributed to feathers and claws (Sánchez-Virosta et al. 2015). To counter the effects of exposure to As, organisms have biotransformation mechanisms that decrease its toxicity, in which inactive As metabolites are formed (monomethylarsenic and dimethylarsenic), which are more easily removed by the kidneys (Soria et al. 1995; ATSDR 2007). In ducklings, clinico-pathological effects caused by sodium arsenate at 30–300 μg/g can produce liver congestion, necrosis and fibrosis, severe degeneration of brain, and increase mortality (Khan et al. 2014).

In general, the levels of As reported in feathers, blood, and organs of penguins are below 3 μg/g, the limit considered normal in living organisms (Jerez et al. 2013a), except for the concentrations of As in blood of little penguins that inhabit St Kilda, on the coast of Australia (3.72 μg/g, Table 6). All the studies performed in penguins reveal that the concentrations of As are below 50 μg/g known as of toxicological significance, or that can lead to endocrine disorders (Neff 1997).

3.1.3 Cadmium

This metal is known to bioaccumulate in marine biota from both natural and anthropogenic sources (Espejo et al. 2014). The maximum concentrations of Cd (351.8 μg/g) have been found in kidneys of Adélie penguins from Avian Island, Antarctic Peninsula (Table 4). In contrast, the lowest concentration of Cd (<0.001 μg/g) has been reported in bones (Table 3) and muscles (Table 6) of chinstrap penguins and Adélie penguins from Antarctica, respectively. In general, birds accumulate Cd in their bodies through the food chain, and Cd is first accumulated in the liver and then transported to several organs (Lee 1996). Cadmium concentrations in penguins tended to be higher in kidneys than in the liver, as also noted in different species of Anseriformes (Jin et al. 2012).

In feathers, the maximum Cd concentrations have been found in adult gentoo penguins from Livingston Island, and the minimum in juvenile Adélie penguins from King George Island (0.01–0.50 μg/g, Table 1). In general, Cd concentrations in penguin feathers are lower than those found in seabirds of the Northern Hemisphere (0.04–1.28 μg/g) (Kim et al. 1998; Agusa et al. 2005; Mansouri et al. 2012). In eggshells, there is little information on the levels of Cd in penguins (Table 2).

In bones, Cd concentrations (<0.001–0.17 μg/g, Table 3) are maximum in Adélie penguins from Avian Island, and minimum in chinstrap penguins from King George Island. The concentrations of Cd in penguin bones are lower than those reported in bones of seabirds of the Northern Hemisphere (0.03–0.33 μg/g, Kim et al. 1998).

In kidneys, Cd concentrations (0.2–351.8 μg/g, Table 4) are highest in Adélie penguins from Avian Island, and are lowest in gentoo and Adélie penguins, both from King George Island. Cadmium levels in kidneys of gulls (0.90–44.4 μg/g) of south-western Poland and the Artic (Orłowski et al. 2007; Malinga et al. 2010) are within the range reported in penguin kidneys. A study found that Cd levels in kidneys of scoters (*Melanitta perspicillata*) from the Queen Charlotte Islands in Canada were as high as 390.2 μg/g, a concentration potentially associated with renal damage (Barjaktarovic et al. 2002).

In the liver, Cd concentrations (0.06–27.7 μg/g, Table 5) are highest in Emperor penguins from the Weddell Sea, and are lowest in Adélie penguins from King

George Island. Cadmium concentrations in penguin livers reveal that 9 of 19 reports (47.4%) exceeded the threshold levels of toxicity for wild birds (3 μg/g, Scheuhammer 1987). The Cd levels in penguin livers are comparable with those levels (0.05–15.1 μg/g) found in seabirds of the Northern Hemisphere (Elliot et al. 1992; Kim and Koo 2007; Pérez-López et al. 2005).

In muscles, Cd levels of seabirds (0.26–0.52 μg/g) from the Northern Hemisphere (Orłowski et al. 2007; Malinga et al. 2010) are within the range reported in penguins (<0.001–2.63 μg/g, Table 6). The highest Cd levels are in Adélie penguins from Avian Island, and are lowest in the same species from Potter Cove.

In the stomach, Cd concentrations (0.09–2.9 μg/g, Table 7) are highest in Adélie penguins from Edmonson Point, and are lowest in gentoo penguins from King George Island. The levels of Cd in penguin stomach contents are far below those levels of Cd detected in the stomach contents of seabirds from industrialized areas of Korea (96–217 μg/g) (Kim and Oh 2014b, c).

In excreta, Cd levels are linked to high dietary Cd intake (Ancora et al. 2002). Cd concentrations (0.16–47.7 μg/g, Table 8) are highest in Humboldt penguins from Pan de Azúcar Island (Chile), and are lowest in Adélie penguins from the Antarctic Peninsula. Levels of Cd in penguin excreta are higher than those observed in wild bird species (0.12–1.88 μg/g) of the Northern Hemisphere (Kaur and Dhanju 2013; Kler et al. 2014).

In birds, the accumulation of Cd can have adverse effects on health, such as renal and testicular damage, disorder in the balance of Ca and the skeletal integrity, reduced feed intake and growth rate, decreased egg laying, thinning eggshells, or alteration in the behavior of the bird, among other effects (Burger 2008; Furness 1996; Larison et al. 2000; Rodrigue et al. 2007). However, seabirds seem to be less vulnerable to the exposure to high levels of Cd than other wild organisms and birds of terrestrial environments (Burger 2008; Furness 1996). Highest Cd concentrations in tissues of marine birds were in kidney tissue of oceanic birds (Elliot et al. 1992; Pérez-López et al. 2005; Orłowski et al. 2007; Kim and Koo 2007; Malinga et al. 2010). In *Pygoscelis* penguins from the South Shetland Islands, a ratio kidney/liver for Cd concentrations of about 4 means a higher Cd affinity for renal tissue (Jerez et al. 2013b), thus indicating a chronic or sub-chronic exposure to Cd due to maternal transfer of this metal during egg development, as occurs in other seabirds (Agusa et al. 2005). A high exposure to Cd causes significant accumulation of this metal in the soft tissues, because a small proportion is excreted, and release of Cd from kidney is very slow (Eisler 1985). Thus, under conditions of chronic dietary exposure, kidney concentrations of Cd may express long-term accumulation (Scheuhammer 1987).

Toxic effects of Cd appear in humans and other mammals when kidney Cd levels reach about 100 μg/g ww (Scheuhammer 1987) or about 400 μg/g dw (assuming a moisture content of 75% in the sample). Seabirds accumulate a large amount of metals such as Hg in their liver because they usually occupy the highest trophic positions in the marine food web and have a long life span (Thompson 1990). However, birds are relatively resistant to some metals, like Cu (Eisler 1998). The process of the metal detoxification in livers of seabirds is well described by Ikemoto et al. (2004). In penguins, some metals interact with others to activate certain phase I detoxification mechanisms in the organism. A study carried out by Kehrig et al. (2015) evidenced a correlation between Se and metallothioneins in liver samples of Magellanic penguins (*Spheniscus magellanicus*), indicating that Se would be involved in detoxification of Cd, Pb, and Hg. Another study showed a positive correlation between Se and Cd in tissues of chinstrap, gentoo, and Adélie penguins, which would be related to the detoxifying function played by Se against the toxicity of Cd (Jerez et al. 2011). In this sense, Jerez et al. (2013a) stated that high levels of Se (30.6 μg/g) and Zn (126.05 μg/g) can protect chinstrap penguins of Deception Island at least partially against high Cd levels (27.54 μg/g) of toxicological significance. However, the accumulation of Cd and Se, and likely other heavy metals, can cause teratogenic effects in a wide range of birds and animal species (Hoffman 2002; Gilani and Alibhai 1990; Ohlendorf et al. 1988; Franson et al. 2007), and even micromelia in penguins (Raidal et al. 2006). High Se levels of over 10 μg/g in liver of aquatic birds can produce hepatic toxicity (Lemley 1993). A study found that 47% of the samples of livers of penguins from Antarctic Peninsula had Se levels above the mentioned toxicity threshold (Jerez et al. 2013a). However, when evaluating Se toxicity and oxidative stress, nutritional factors should be taken into consideration (Franson et al. 2007).

Studies carried out in colonies of some penguins from Antarctica have shown that kidney samples collected at Weddell sea and Avian Island present high concentrations of Cd (270.2 and 351.8 μg/g, Table 4), implying that those seabirds probably presented a chronic exposure to this metal, with levels above the toxicity threshold established for birds (Furness 1996).

3.1.4 Mercury

The maximum concentrations of Hg (8.16 μg/g, Table 1) have been found in feathers of adult gentoo penguins from Crozet Islands. In contrast, the lowest concentration of Hg (0.005 μg/g) has been reported in eggshells of Adélie penguins from Admiralty Bay, Antarctica (Table 2). As with most seabirds, penguin feathers constitute an important way of detoxification of Hg (Yin et al. 2008).

In feathers, Hg concentrations (0.033–8.16 μg/g, Table 1) are highest in adult gentoo penguins from Crozet Islands (Carravieri et al. 2016). The lowest Hg levels have been reported in juvenile Magellanic penguins from the coasts of Argentina (Frias et al. 2012). Mercury concentrations in penguin feathers are lower than those found in different species of seagulls and terns from Northern Hemisphere (0.31–20.2 μg/g) (Goutner et al. 2000; Zamani-Ahmadmahmoodi et al. 2014) and in feathers of birds from various locations of the Chilean coast (0.11–13 μg/g, Ochoa-Acuña et al. 2002). Of the thirty-two reports in penguin feathers, only two studies are in the range of Hg levels (5–40 μg/g) linked to reduced hatched of egg laid in various bird species (Eisler 1987). Concentrations of Hg of 9–20 μg/g in feathers can decrease reproductive success in some piscivorous birds (Fimreite

1974; Scheuhammer 1987; Beyer et al. 1997; Evers et al. 2008). The range of Hg concentrations reported in penguin feathers are below those known to cause adverse health and reproductive effects in birds.

In eggshells, Hg concentrations (0.005–0.26 μg/g, Table 2) are highest in Adélie penguins from Terra Nova Bay (Bargagli et al. 1998), and are lowest in the same species in Almiralty Bay (Santos et al. 2006). Mercury levels in penguin eggshells are lower than those reported in marine, aquatic, and terrestrial birds of other latitudes (0.05–36.37 μg/g) (Yin et al. 2008; Daso et al. 2015).

In bones, a single study reported Hg concentrations (0.02 μg/g, Table 3) in Adélie penguins that inhabit the surroundings of the Zhongshan Station (Yin et al. 2008). In general, data of Hg in bones of birds are not abundant, because this metal is not precisely stored in this biotic matrix, making comparisons difficult. In any case, levels of Hg in penguin bones are 50% lower than those detected in bones of seagulls from the coasts of Japan (Agusa et al. 2005) and much lower than those in great cormorants (*Phalacrocorax carbo*) from Europe (1.4–1.72 μg/g, Skoric et al. 2012).

In kidneys, Hg concentrations (0.146–2.47 μg/g, Table 4) are highest in Magellanic penguins from the coast of Southern Brazil (Kehrig et al. 2015). The lowest levels are reported in Adélie penguins that inhabit King George Island (Smichowski et al. 2006). Mercury concentrations in penguin kidneys are lower than those detected in kidneys of seabirds from the Northern Hemisphere (0.3–5 μg/g) (Arcos et al. 2002; Zamani-Ahmadmahmoodi et al. 2014).

In livers, Hg concentrations (0.269–5.7 μg/g, Table 5) are highest in Magellanic penguins from the coasts of Southern Brazil (Kehrig et al. 2015). The lowest concentrations of Hg have been reported in Adélie penguins from Potter Cove, King George Island (Smichowski et al. 2006). Mercury concentrations in penguin livers are below those reported in seabirds of the Northern Hemisphere (4.9–306 μg/g, Kim et al. 1996). In birds, sublethal effects of Hg on growth, development, reproduction, blood and tissue chemistry, metabolism, behavior, histopathology, and bioaccumulation have been found between 4 and 40 mg/kg (dietary intake) (Eisler 1987). The concentrations of Hg in liver of Magellanic penguins from Rio Grande do Sul, Brazil (5.7 μg/g, Table 5) are higher than the threshold of toxicity for Hg (Kehrig et al. 2015).

In muscles, Hg is reported by a single study in Adélie penguins (0.6 μg/g, Table 6) from Terra Nova Bay (Bargagli et al. 1998). Levels of Hg in penguin muscles are lower than those reported in terns and gulls from Asia (0.9–2.5 μg/g, Zamani-Ahmadmahmoodi et al. 2014).

In stomachs, Hg (0.08–0.10 μg/g, Table 7) is lowest in Adélie penguins from Terra Nova Bay (Bargagli et al. 1998) and is highest in the same species from Edmonson Point (Ancora et al. 2002). It was difficult to find more reports of Hg levels in bird stomachs. Levels of Hg detected in penguin stomachs are much lower than those measured in intestines of cormorants from Europe (1.29–2.49 μg/g, Skoric et al. 2012).

In excreta, Hg concentrations (0.06–6.60 μg/g, Table 8) are highest in gentoo penguins from O'Higgins Base, and are lowest in chinstrap penguins from Barton

Peninsula, both locations of the Antarctic Peninsula. Levels of Hg in penguin excreta are higher than those in other marine birds worldwide (0.10–0.75 μg/g, Yin et al. 2008).

Mercury concentrations in penguin blood (0.84–2.75 μg/g) and in penguin brains (0.43 μg/g) are been measured in little penguins from Australia and Adélie penguins from Terra Nova Bay (Antarctica), respectively (Table 9). Those levels are higher than those found in the blood of black-tailed gull chicks and Great tits from the Northern Hemisphere (0.03–0.26 μg/g) (Dauwe et al. 2000; Kim et al. 2013). Mercury concentrations of over 3 μg/g in blood can affect endocrine systems of Arctic birds with negative consequences for reproduction (Tartu et al. 2013). In loons (*Gavia immer*), Evers et al. (2008) reported an adverse effect threshold for adult birds at 3 μg/g (w.w) in blood and reproductive failure when adult blood Hg levels reach 12 μg/g/(w.w). Tartu et al. (2016) found that Hg levels (1.0–1.5 μg/g) in blood of adult kittiwakes can disrupt prolactin secretion (a pituitary hormone involved in parental care) which could lead to reduced chick survival.

Chronic exposure to metals may imply a threat to penguins. Some evidence shows that the survival and breeding success decreased with increasing Hg levels in blood of Artic seabird (2.28 ± 0.42 μg/g, Goutte et al. 2015). Mercury in its organic form (methylmercury, ethylmercury) is more lipophilic, which favors its accumulation mainly in the liver, kidneys, brain, and feathers. Inorganic Hg is mostly accumulated in kidneys, due to its affinity to metallothioneins presented by renal cells (Byrns and Penning 2011). In seabirds, habitat type and functional feeding group may influence organic Hg bioaccumulation rates at higher trophic levels (Chen et al. 2008). The direct effects of elevated organic Hg on marine biota can include changes in brain neurochemical receptor density (Scheuhammer et al. 2008). In pinnipeds, adverse effects may manifest as immunosuppression (Lalancette et al. 2003). There are few studies on the effects of metals in feathers and blood of birds, but evidence exits indicating that concentrations of Hg of 5 μg/g in feathers of birds can cause reproductive impairment (Burger and Gochfeld 1997), including smaller egg size, lower hatching rate, decreased chick survival, and even impaired territorial fidelity in waterfowl (Rothschild and Duffy 2005). The few studies that exist reveal that the concentrations of Hg in biotic matrices of penguins from Antarctica are below the stated threshold of toxicological significance for Hg. In general, Hg levels are lower in most of the biological matrices of penguins than birds from the Northern Hemisphere.

3.1.5 Lead

Excepting excreta, the maximum concentrations of Pb (almost 1.90 μg/g) have been found in feathers of adult gentoo penguins from Livingston Island (Table 1) and in bones of Adélie penguins from East Antarctica (1.60 μg/g, Table 3). In contrast, the lowest concentration of Pb (<0.001 μg/g) has been reported in kidneys (Table 4), liver (Table 5), and muscles (Table 6) of gentoo penguins from King George Island.

Lead is not metabolically regulated (Gochfeld et al. 1996), and unlike Cd, tends to be accumulated in bird feathers (Jerez et al. 2011).

In feathers, Pb concentrations (Table 1) are highest (almost 1.90 μg/g) in adult gentoo penguins from Livingston Island. On the other hand, Pb levels are lowest (<0.01 μg/g) in juvenile Adélie penguins from King George Island. The highest concentration of Pb in penguin feathers is directly related to major human activity (Jerez et al. 2011, 2013a). Levels of Pb in penguin feathers are lower than those concentrations found in feathers (0.34–7.15 μg/g) of different seabirds of the Northern Hemisphere (Kim et al. 1998; Burger et al. 2008; Ribeiro et al. 2009; Skoric et al. 2012; Kim and Oh 2014b). Lead concentrations of 4 μg/g (dw) in feathers are known to be a threshold level for toxicity (Burger and Gochfeld 2000b).

In eggshells, Pb levels in penguin eggshells are rare. The highest Pb concentrations (0.75 μg/g, Table 2) have been found in gentoo penguins from Fildes Peninsula (Yin et al. 2008). Levels of Pb (0.68–0.75 μg/g) in eggshells of penguins are lower than those reported in seabirds (1.25–3.10 μg/g) of the Northern Hemisphere (Yin et al. 2008; Kim and Oh 2014a).

In bones, Pb concentrations (<0.001–1.60 μg/g, Table 3) are highest in Adélie penguins from Zhongshan Station (Yin et al. 2008), and are lowest in *Pygoscelis* penguins from King George Island and Byers Peninsula (Barbosa et al. 2013; Jerez et al. 2013a). The concentrations of Pb in bones of penguins are lower than those reported in bones of marine, aquatic, and terrestrial bones (0.04–42.32 μg/g) of the Northern Hemisphere (Lebedeva 1997; Kim et al. 1998; Orłowski et al. 2007; Yin et al. 2008). Lead is known to be a toxic metal, and the skeleton is the main depot for these elements (Lebedeva 1997). Lead levels >10 μg/g in bone of wild birds are considered to be toxic, and so may be interpreted as a result of relatively polluted habitats (Scheuhammer 1987). Bone Pb concentrations higher than 20 μg/g are considered as excessive exposure for waterfowls (Franson 1996). Levels in penguin bones are far below those threshold values, which suggest that the biological effect should be neglected.

In kidneys, Pb concentrations (<0.001–0.55 μg/g, Table 4) are highest in Magellanic penguins from the coast of Southern Brazil (Kehrig et al. 2015), and are lowest in gentoo penguins from King George Island (Jerez et al. 2013b). Concentrations of Pb in penguin kidneys are lower than those of seabirds from the Northern Hemisphere (0.14–11.18 μg/g) (Kim et al. 1998; Orłowski et al. 2007). Lead concentrations above 68 μg/g in kidneys of snowy owls (*Nyctea scandiaca*) are linked to bird's mortality (Franson 1996).

In the liver, Pb levels varies from <0.001 to 0.58 μg/g (Table 5) with the highest concentrations in Magellanic penguins from the coasts of Southern Brazil (Kehrig et al. 2015). The lowest levels are reported in gentoo penguins from King George Island (Jerez et al. 2013b). Concentrations of Pb in penguin livers are lower than values (0.50–3.71 μg/g) found in seabirds of Asia (Kim et al. 1998; Kim and Koo 2007; Kim and Oh 2014c). A study conducted in South Korea (Kim and Oh 2014c) found that high levels of Pb in liver (6.2 μg/g) could negatively affect both behavior and growth of chicks of the black-tailed gull. Concentrations of Pb in livers of penguins are far below this threshold value. Hepatic Pb concentrations of over

30 μg/g in waterfowls can produce Pb poisoning, which is characterized by impaction of the upper alimentary tract, submandibular edema, myocardial necrosis, and biliary discoloration of the liver (Beyer et al. 1998).

In muscles, Pb concentrations (Table 6) are highest (almost 0.60 μg/g) in gentoo penguins from Livingston Island (Metcheva et al. 2010), and are lowest (<0.001 μg/g) in the same species of King George Island (Jerez et al. 2013b). In general, the levels of Pb in penguin muscles are lower than those reported in seabirds (0.014–3.59 μg/g) of the Northern Hemisphere (Kim et al. 1998; Orłowski et al. 2007).

In the stomach, Pb concentrations (0.03–0.71 μg/g, Table 7) are highest in gentoo penguins from King George Island, and are lowest in chinstrap penguins from King George Island. The levels of Pb in stomach contents of penguins are lower than those levels of Pb (0.059–105.0 μg/g) detected in stomach contents of seabirds from the Northern Hemisphere (Kim et al. 1998; Kim and Oh 2014b, c).

In excreta, Pb concentrations (0.08–12.79 μg/g, Table 8) are highest in Humboldt penguins from Cachagua Island (Celis et al. 2014), while the lowest levels were reported in gentoo penguins from Neko Harbor, Antarctic Peninsula (Celis et al. 2015b). In general, levels of Pb in penguin guano are lower than the concentrations of Pb (3.90–124.8 μg/g) in guano of aquatic and terrestrial birds from the Northern Hemisphere (Dauwe et al. 2000; Martinez-Haro et al. 2010; Kler et al. 2014).

In blood, Pb concentrations (0.04–0.07 μg/g, Table 9) have been measured only in little penguins from the coast of Australia. Those Pb levels are below the deleterious effect level of 4 μg/g (Finger et al. 2015), and are also lower than those reported in gulls from the Northern Hemisphere (0.06–0.18 μg/g, Kim et al. 2013). Some biological functions of birds can be altered when Pb levels in blood >3 μg/g and Pb levels >6 μg/g can produce uremic poisoning (Franson 1996).

In birds, it has been observed that the exposure to Pb in young individuals of the herring gull (*Larus argenteus*) and the common tern (*Sterna hirundo*) affects behavioral development, growth, locomotion, balance, search for food, thermoregulation, and recognition between individuals (Burger and Gochfeld 2000a). Pb is transported through blood bonded to hemoglobin, reaching the liver, kidneys, bone marrow, and central nervous system. Nevertheless, Pb can be stored in tissues rich in Ca such as hairs, feathers, and bones, where it can remain for many years (O'Flaherty 1998). Lead in penguin bones is accumulated throughout the lifetime of the individual, and so its presence in bones may be considered an indicator of long-term exposure (Barbosa et al. 2013). A study for *Pygoscelis* penguins from Antarctica found that Cd, Ni, Pb, and Se levels in muscles are long-term dependent (Jerez et al. 2013a). High concentrations of Cu might increase the effects of toxicological significance in penguins caused by Pb (Eisler 1988).

Feces can be used to detect adverse toxicological effects in wildlife by means of porphyrins, which can be correlated with metals measured in the same sample (Mateo et al. 2016). A study showed a strong affinity between the levels of Pb with porphyrins in excreta of gentoo penguins (Celis et al. 2012), which may be associated with hepatic and renal damage (Casini et al. 2003). Available data indicate that concentrations of Pb in guano of penguins in the Antarctica have increased in the last 200 years as a result of greater local anthropogenic activity (Sun and Xie 2001). Studies that are able to show the possible biological effects of Pb on these populations of polar seabirds are needed.

Negative correlations between Pb–Cu and Pb–Fe have been found in livers of *Pygoscelis* penguins (Jerez et al. 2013a), indicating the capability of Pb (a metal directly linked to various anthropogenic activities) to use the transport mechanisms of the essential cations, preventing them from performing their metabolic function (Ballatori 2002). Penguin species from higher latitudes could be more vulnerable to the effects of trace elements due to their less effective immunological systems in such environments in comparison to other species of penguins that live in lower latitudes (Boersma 2008; Cooper et al. 2009).

There are few studies on the exposure to heavy metals in penguins and it is necessary to progress in the use of non-destructive biomarkers and non-invasive matrices (i.e., feathers or fecal material) or semi-invasive such as blood tissue. Porphyrins have proved to be useful biomarkers of exposure to contaminants (Casini et al. 2003), because they are capable of bonding to metals and they can be detected in different biological matrices (De Matteis and Lim 1994). Some trace metals can interfere with the biosynthesis of hemoglobin and cause alterations in the porphyrins, which are accumulated or excreted (Casini et al. 2001). Byproducts such as copro- uro- and protoporphyrins are not toxic in normal concentrations, but when there is an excess they can affect the liver and bone marrow (Lim 1991). A study showed a positive correlation between the levels of porphyrins and those of Hg and Pb in guano of gentoo penguins (Celis et al. 2012). Another study carried out in Humboldt penguins found that the levels of porphyrins were directly correlated with the concentrations of As, Pb, and Cu, thus there exists a high probability that these penguins might develop hepatic and renal damage because of the exposure to these metals (Celis et al. 2014). The higher concentrations of metals in penguin excreta suggest a physiological mechanism of detoxification (Ancora et al. 2002), although this may also imply that those trace elements are not absorbed at the intestinal level. It has been observed that when birds present renal damage caused by Cd, the levels of this metal in excreta are increased (Goyer 1997). Lead concentrations in all of the biotic matrices of penguins studied are lower than those Pb levels found in marine, aquatic, and terrestrial birds of the Northern Hemisphere, which is highly industrialized and where human population is concentrated.

3.2 Essential Trace Elements

3.2.1 Copper

In general, there are not enough data available on the toxicity of Cu to avian wildlife. Birds, when compared to lower forms (fish, amphibians) are relatively resistant to Cu (Eisler 1998). With the exception of excreta, the maximum concentrations of Cu have been found in the liver of gentoo penguins from King George Island (386.1 μg/g, Table 5). In contrast, the lowest concentration of Cu (0.06 μg/g) has been reported in bones of Adélie penguins from the same location (Table 3). There is evidence showing that Cu levels of 1050 μg/g in the livers of eiders can cause liver necrosis and fibrosis (Norheim and Borch-Iohnsen 1990). In pygoscelid penguins, Cu levels over 24 μg/g in the liver (Szefer et al. 1993) could represent an additional stress to birds already facing stressful conditions, such as starvation (Debacker et al. 2000).

In feathers, Cu concentrations (6.87–20.89 μg/g, Table 1) are highest in adult gentoo penguins from O'Higgins Base (Celis et al. 2015b), whereas they are lowest in juveniles of the same species from King George Island (Jerez et al. 2013b). Levels of Cu in penguin feathers are higher than those concentrations found in feathers of different seabirds (7.56–11.2 μg/g) of the Northern Hemisphere (Kim et al. 1998; Malinga et al. 2010).

In eggshells, Cu concentrations in penguins are scarce and there is a single study (1.24 ± 0.4 μg/g, Table 2) in gentoo penguins from Livingston Island (Metcheva et al. 2011). Copper concentrations in penguin eggshells are comparable to those Cu levels reported in eggshells of birds from other latitudes (0.42–7.54 μg/g) (Dauwe et al. 2000; Yin et al. 2008; Kim and Oh 2014a).

In bones, Cu concentrations (0.06–1.15 μg/g, Table 3) are highest in colonies of Gentoo penguins from Byers Peninsula (South Shetland Islands). Concentrations of Cu in penguin bones are lower than those Cu levels found in bones of marine, aquatic, and terrestrial birds of the Northern Hemisphere (0.37–60 μg/g) (Lebedeva 1997; Kim et al. 1998; Orłowski et al. 2007; Yin et al. 2008).

In kidneys, Cu has been reported between 1.6 and 19.99 μg/g (Table 4), with the highest concentrations in Gentoo penguins from King George Island, whereas the lowest levels correspond to Adélie penguin from Potter Cove (King George Island). Levels of Cu in penguin kidneys are lower than those found in kidneys of Artic seabirds (12.2–27.8 μg/g) (Kim et al. 1998; Malinga et al. 2010).

In livers, Cu concentrations (10.91–386.1 μg/g, Table 5) are highest in colonies of gentoo penguins from King George Island, and are lowest in Adélie penguins from the same location. The levels of Cu in livers of Antarctic penguins are higher than those detected in other seabirds of Asia and Europe (0.26–92.5 μg/g) (Kim and Koo 2007; Pérez-López 2005; Ribeiro et al. 2009; Malinga et al. 2010). A study found that mute swans (*Cygnus olor*) from estuaries in Britain had more than 2000 μg/g of Cu in their blackened livers (Bryan and Langston 1992).

In muscles, Cu concentrations (4.43–9.95 μg/g, Table 6) are highest in colonies of gentoo penguins from King George Island (Jerez et al. 2013a), whereas they are lowest in Adélie penguins from King George Island (Jerez et al. 2013b). Levels of Cu in penguin muscles are within the range reported in the muscles of seabirds from Northern Hemisphere (3.59–18.3 μg/g) (Kim et al. 1998; Malinga et al. 2010; Orłowski et al. 2007).

In stomachs, Cu levels (4.85–66.42 μg/g, Table 7) presented the highest value in Adélie penguins from Avian Island, and the lowest levels in the same species from King George Island. The levels of Cu in penguin stomach contents are higher than those detected in seabirds of the Northern Hemisphere (4.89–14.0 μg/g) (Kim et al. 1998; Kim and Oh 2014b).

In excreta, Cu concentrations (37.6–585.8 μg/g) are highest in colonies of Adélie penguins from Kopaitic Island, and are lowest in chinstrap penguin from the Antarctic Peninsula (Table 8). Levels of Cu in penguin guano are higher than those values (10–150.8 μg/g) found in excrement birds from other parts of the world (Dauwe et al. 2000; Yin et al. 2008; Kler et al. 2014). A study in excreta of Humboldt penguins found that the levels of porphyrins were directly correlated with the concentrations of As, Pb, and Cu (Celis et al. 2014), and those birds might present some hepatic and renal disorder (Casini et al. 2003).

In blood, Cu concentrations (2.14–2.48 μg/g, Table 9) are only reported in little penguins from Australia. Copper concentrations in penguin blood are within the range reported in the seagulls, eiders, and ducks of the Northern Hemisphere (0.64–2.56 μg/g) (Franson et al. 2003; Kim et al. 2013).

In general, marine birds retain a very small portion of Cu and other metals ingested (Bryan and Langston 1992). Although Cu is an essential metal, in excess it can produce a series of metabolic, pulmonary, hepatic, and renal toxic effects (Soria et al. 1995). Copper can increase the toxic effects caused by Pb in birds, fishes, and invertebrates (Eisler 1988). In birds, Cu is accumulated in the liver and bone marrow, being associated with metallothionein and thus preventing an excess of free ions of this element (Eisler 1998). However, this protective mechanism is limited and lesions can be produced in the liver (ATSDR 2004).

3.2.2 Manganese

Excepting excreta and stomach contents, the maximum concentrations of Mn have been found in bones (18.35 μg/g, Table 3) and the liver (15.83 μg/g, Table 5) of gentoo penguins from Byers Peninsula and Adélie penguins from King George Island, respectively. In contrast, the lowest concentration of Mn (<0.01 μg/g) has been reported in feathers of juvenile Adélie penguins from Avian Island, Antarctica (Table 1).

In feathers, Mn concentrations range <0.01–3.26 μg/g (Table 1), with the highest levels in chinstrap penguins at Deception Island and the lowest in Adélie penguins at Avian Island. This range in penguin feathers is lower than those found in seabirds from the Northern Hemisphere (0.003–19.29 μg/g) (Burger et al. 2008; Kim et al. 2013), and also is lower than Mn levels detected in feathers of adult seabirds from industrialized and populated areas, such as the Brazilian coasts (11.4 μg/g, Barbieri et al. 2010).

In eggshells, there is only one study reporting concentrations of Mn (0.82 ± 0.08 μg/g, Table 2) in gentoo penguins from Livingston Island (Metcheva et al. 2011). Manganese concentration in penguin eggshells is within the range reported in the seabirds of the United States and Spain (0.29–4.63 μg/g) (Gochfeld 1997; Morera et al. 1997).

In bones, Mn concentrations (2.5–18.35 μg/g, Table 3) are highest in gentoo penguins from Byers Peninsula (Barbosa et al. 2013), and are lowest in the same species from Livingston Island (Metcheva et al. 2010). Concentrations of Mn in bones of penguins are within the range reported in the bones of marine, aquatic, and terrestrial birds of the Northern Hemisphere (1.06–30.6 μg/g) (Lebedeva et al. 1997; Kim et al. 1998).

In kidneys, Mn concentrations (3.78–11.18 μg/g, Table 4) are highest in chicks of Adélie penguins from King George Island during the 2008–2009 austral summer season, and are lowest in adult Adélie penguins from the same location during austral summers of 2007–2010. Manganese concentrations are higher in penguin chicks than those of adult specimens. Although Mn levels detected in individuals of the same species seem to show a temporal variability, the age of the birds seems to be relevant; birds regulate Mn primarily by excretion in the feces (Kler et al. 2014), and probably Mn intake from food in chicks exceeds excretion (Skoric et al. 2012). Concentrations of Mn in penguin kidneys are within the range found in the kidneys of Arctic seagulls (<0.01–20.1 μg/g; Malinga et al. 2010).

In the liver, Mn concentrations (6.8–15.83 μg/g, Table 5) are highest in Adélie penguins from King George Island and are lowest in the same species from the Antarctic Peninsula. Levels of Mn in penguin livers are comparable to values reported in seabirds from Asia and Artic (4.14–20.3 μg/g) (Kim et al. 1998; Malinga et al. 2010).

In muscles, Mn concentrations (0.46–2.55 μg/g, Table 6) are highest in chinstrap penguins from Deception Island and are lowest in gentoo penguins from the Antarctic Peninsula. Concentrations of Mn in penguin muscles are slightly lower than the concentrations of Mn in muscle tissues of Arctic birds (1.84–2.56 μg/g) (Campbell et al. 2005; Burger et al. 2008).

In the stomach, Mn levels (2.20–82.43 μg/g, Table 7) are highest in gentoo penguins from King George Island and are lowest in Adélie penguins from Avian Island. The levels of Mn in penguin stomach contents are higher than those of Mn (0.98–15.9 μg/g) detected in stomach contents of seabirds from the Northern Hemisphere (Kim et al. 1998; Kim and Koo 2007).

In excreta, Mn levels (12.3–138 μg/g, Table 8) are highest in chinstrap penguins from the Antarctic Peninsula and are lowest in gentoo penguins from Livingston Island. Concentrations of Mn in penguin droppings are within the range (0.03–221.8 μg/g) found in the guano of different avian species from Asia (Lebedeva et al. 1997; Kaur and Dhanju 2013; Kler et al. 2014).

In animals, Mn is a neurotoxic metal that can affect several neural activities, and at concentrations of about 1000 μg/g, it has negative effects on certain brain functions (Šaric and Lucchini 2007). Mn is distributed via blood linked to proteins (eg. albumin), being accumulated in tissues rich in mitochondria, such as hepatic and renal tissue (Erikson and Aschner 2003; Soria et al. 1995). Effects produced by an acute exposure to Mn include irritation in the digestive tract, respiratory disorders, cardiac ailments, coma, and even death (Soria et al. 1995). In turn, chronic intoxication with this metal generates neurological, reproductive, pulmonary, and immune alterations (ATSDR 2008). The elimination of Mn is produced mainly through the gastrointestinal tract (Roth 2006).

No research has been done related to the effects of Mn on penguins. It is an issue because increases in the environmental Mn levels have been related to the current use of Mn as additive in combustibles (Burger and Gochfeld 2000b). There is recent evidence showing that Mn levels in hepatic tissues of Antarctic penguins (Jerez et al. 2013a) are slightly higher than those detected two decades ago (Honda et al. 1986; Szefer et al. 1993).

3.2.3 Zinc

With the exception of excreta, the maximum concentrations of Zn (330.3 μg/g) have been found in livers of chinstrap penguins from King George Island (Table 5) and in bones of the same species from Byers Peninsula, Antarctica (Table 3). In contrast, the lowest concentration of Zn (4.07 μg/g) has been reported in eggshells of gentoo penguins from Livingston Island, Antarctica (Table 2).

In feathers, the range of Zn concentrations (33.26–119.72 μg/g, Table 1) indicates that the highest concentrations are in adult gentoo penguins from King George Island (Jerez et al. 2013a) and the lowest are in juvenile same species from Doumer Island. Zn levels in penguin feathers are similar to those Zn levels found in feathers of different seabirds of the Northern Hemisphere (42.9–189.2 μg/g) (Kim et al. 1998, 2013; Ribeiro et al. 2009; Lucia et al. 2010).

In eggshells, studies on Zn in penguins are not abundant. Zinc concentrations (4.07–8.3 μg/g, Table 2) are highest in Adélie penguins from Admiralty Bay, and are lowest in gentoo penguins from Livingston Island, South Shetland Islands. The levels of Zn in penguin eggshells are lower than those detected in water birds and seabirds of the United States and the Artic (9.04–58.1 μg/g) (Custer et al. 2007; Malinga et al. 2010).

In bones, the range of Zn (81–244.6 μg/g, Table 3) indicates the highest concentrations are in gentoo penguins from Byers Peninsula, whereas the lowest concentrations are in the same species from Livingston Island. The concentrations of Zn in penguin bones are similar to those Zn levels reported in bones of marine birds of the Northern Hemisphere (83.9–202 μg/g) (Kim et al. 1998; Yin et al. 2008; Skoric et al. 2012).

In kidneys, Zn concentrations (85.74–234.3 μg/g, Table 4) are highest in Adélie penguins from Avian Island. In contrast, the lowest Zn concentrations are in the same species from King George Island. Levels of Zn in penguin kidneys are higher than those Zn levels found in kidneys of marine birds from the North Pacific and Artic seabirds (30.2–183 μg/g) (Kim et al. 1998; Sagerup et al. 2009; Malinga et al. 2010).

In the liver, Zn concentrations (72–330.34 μg/g, Table 5) are highest in chinstrap penguins from King George Island, and are lowest in gentoo penguins from Livingston Island. Concentrations of Zn in penguin livers are above those Zn levels found in seabirds of the Northern Hemisphere (14.92–541 μg/g) (Parslow et al. 1973; Kim and Koo 2007; Pérez-López et al. 2005). A study found that a high concentration of Zn (541 μg/g) in livers of northern gannets (*Morus bassanus*) could be the main cause of the bird's mortality (Parslow et al. 1973).

In muscles, Zn concentrations (24–163.75 μg/g, Table 6) indicate the highest concentrations are in Adélie penguins from King George Island, while the lowest concentrations are in gentoo penguins from Livingston Island. Levels of Zn in seabirds (53.2–75.5 μg/g) of the Northern Hemisphere (Kim et al. 1998; Malinga et al. 2010) are within the range found in the penguin muscles.

In stomachs, Zn levels (19.84–71.16 μg/g, Table 7) are highest in Adélie penguins from King George Island, and are lowest in gentoo penguins from the same location. Concentrations of Zn in penguin stomach contents are within the range (6.64–102 μg/g) found in the seabirds of the Northern Hemisphere (Kim et al. 1998; Kim and Koo 2007).

In excreta, the range of Zn (0.83–487.1 μg/g, Table 8) shows the highest concentrations in Humboldt penguins from Pan de Azúcar Island, while the lowest levels are in the same species from Cachagua Island. Concentrations of Zn in penguin droppings are lower than those (100–721.8 μg/g) found in marine birds and different avian species of the Northern Hemisphere (Yin et al. 2008; Kaur and Dhanju 2013).

In blood, only a single study has measured Zn levels in little penguins (33.47–38.77 μg/g, Table 9). These levels are within the range detected in the long-tailed ducks (*Clangula hyemalis*) and nesting common eiders (*Somateria mollissima*) from Alaska (18.2–39 μg/g) (Franson et al. 2003).

Despite the fact that Zn is an essential metal, some pancreas histological damage has been detected in birds at high Zn levels (Eisler 1993). In birds, Zn accumulated in liver bonded to metallothionein, though it can also be accumulated in muscles and bones (Wastney et al. 2000). In seabirds, there is a significant positive correlation between renal Zn and Cd, which evidences a possible effect of metallothionein synthesis caused by Cd accumulation (Honda et al. 1990; Malinga et al. 2010). Evidence shows that Zn poisoning in birds usually occurs when the concentration of this metal exceeds 2100 μg/g in the liver or kidney (Eisler 1993). The concentrations of Zn in livers of penguins are below 200 μg/g (Table 5), considered as the threshold value of physiological importance in different species of seabirds (Honda et al. 1990), except that found in liver of chinstrap penguins (330.3 μg/g) and in livers of gentoo penguins (237.2 μg/g) from King George Island. These levels of Zn seem to be related to the great concentration of human activities present in King George Island (Tin et al. 2009).

4 Similarities and Differences of Trace Elements

4.1 Distribution of Trace Elements

There is great similarity (82%) between concentrations of trace elements in guano and stomach contents of penguins (Fig. 2). Likewise, the levels of trace elements in

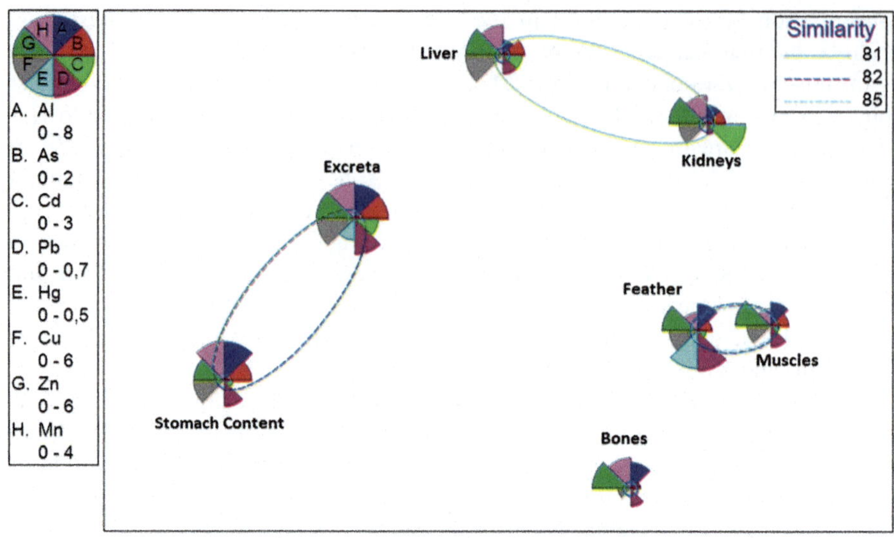

Fig. 2 Bubble chart for mean concentrations of metals in different biological matrices of *Pygoscelis papua* reported from South Shetland Islands. Data taken from Barbosa et al. (2013), Brasso et al. (2014), Celis et al. (2015b), Espejo et al. (2014), Jerez et al. (2011), Jerez et al. (2013a, b), Metcheva et al. (2006, 2010, 2011), Santos et al. (2006), Yin et al. (2008)

kidneys and livers present great similarity (81%), which may be due to the fact that both organs have similar mechanisms of detoxification and biotransformation of elements (Sánchez-Virosta et al. 2015). Furthermore, there is also 85% similarity between the concentrations of trace elements in feathers and muscles. Due to sampling constraints, it is not easy to establish relationships between the concentrations in feathers and the concentrations found in the internal tissues of penguins, even though some previous works have found some relationship between feathers and muscle tissues for some trace elements in birds (Del Hoyo et al. 1992). Metal levels in eggs and bones presented no correlation with the other biotic matrices. Some metals (as Pb, Cd) are not metabolically regulated and tend to be immobilized in bird bones (Lebedeva 1997) and eggshells (Kim and Oh 2014a). Both biological matrices are mainly composed of Ca, which is one of the most important plasma constituents in mammals and birds, and provides structural strength and support to bones and eggshells (De Matos 2008). Trace elements such as Pb and Cd might interact with the metabolic pathway of Ca (Scheuhammer 1987).

A high content of elements in penguin excreta imply a physiological mechanism of detoxification (Ancora et al. 2002), but also imply that elements are not necessarily absorbed at the intestinal level, which reinforces the fact that high concentrations of trace elements in feces are likely the result of low intestinal absorption rather than detoxification mechanisms, and much of the elements ingested by these seabirds are being excreted. It is observed that when some bird present renal damage caused by Cd, the levels of this metal in excreta is increased (Goyer 1997). In penguins, feathers play an important role in detoxification of Hg and Pb, because a large amount of these metals from their diets can be transferred into their plumages (Stewart et al. 1997; Ancora et al. 2002; Jerez et al. 2011). Other metals such as Cd, Cu, and Zn are mainly eliminated via the feces (Ancora et al. 2002; Yin et al. 2008). In general, sequestration of metals (such as Hg or Pb) in bird's feathers results in decreased internal bioavailability (Jerez et al. 2011; Calle et al. 2015). Diet, exposure levels, physiological conditions, and the toxic-kinetic mechanisms regulate the arrival of metals to feathers, as in the case of Hg (Becker et al. 2002). Redistribution to plumage occurs during feather growth when the feather is connected to blood vessels, and metals are incorporated in the keratin structure (Burger et al. 2011). When the feather matures, blood vessels shrivel, and the feather is no longer supplied with blood, at which point metal deposition to the feather ceases (Burger 1993). Hg elimination is possible via deposits in eggs, excreta, uropygial gland, and feathers (Dauwe et al. 2000). In seabirds, Hg concentrations in feathers reflect the uptake and storage of this heavy metal during the period between molts rather than short-term uptake (Furness et al. 1986).

In general, trace element levels in penguins are scarce and fragmented; therefore, no correlation analysis is possible now. Data of trace elements available in *Pygoscelis* penguins of the South Shetland Islands indicate that the levels of Cd in gentoo, chinstrap, and Adélie penguins that live there are strongly influenced by diet, which has also been noted in populations of seagulls from the Northern Hemisphere (Kim and Oh 2014c). In birds, trace element levels in blood reflect recent dietary exposure and often correlate strongly with those in internal tissues (Monteiro and Furness 2001). A study evidenced that blood provides a more precise indicator of penguin body burden for Al, As, Cd, Cu, Fe, Hg, Pb, Se, and Zn than feathers (Finger et al. 2015).

At present, most Hg pollution resides in aquatic environments, where it is converted to methylmercury (Chen et al. 2008). Because of its high affinity with sulfydryl groups of proteins, this heavy metal is easily incorporated into the food chain, bioaccumulating in aquatic organisms, and bioamplifying from one trophic level to the next (Fitzgerald et al. 2007). Some metals such as Zn and Cd among others might be biomagnified under certain environments such as Antarctica, a cold place where trophic chains are short and highly dependent on krill (Majer et al. 2014). The whole trophic transference coefficient (TTC) for gentoo penguins at King George Island is 0.01 for Al, 0.21 for As, 39.55 for Cd, 0.21 for Pb, 1.45 for Cu, 6.90 for Zn, and 0.05 for Mn (no data were available for Hg levels in stomach contents of the species at that location). The value of TTC is usually <1 for trace metals (Anan et al. 2001), except for those metals highly cumulative in the organism which are biomagnified in the trophic chain, such as Hg (Lavoie et al. 2013). Cd and Zn showed a high cumulative power in gentoo penguins, with TTC values far above unity. Scientific evidence indicates that Zn, Se, Cu, and Cd tend to bioaccumulate in aquatic trophic chains (Dehn et al. 2006; Mathews and Fisher 2008). This suggests the possibility of metal biomagnification under specific circumstances. It has been found that biomagnification of Hg is expressed more strongly in cold environments with simple trophic chains (Lavoie et al. 2013). This issue should be addressed in depth in further studies, considering the diversity of marine environments in which the different species of penguins feed.

4.2 Geographical Differences

Most studies of levels of trace elements in penguins (Table 10, Fig. 3) have been carried out in Antarctica and nearby islands. The most reported trace elements are Pb, Cd, Cu, and Zn. In contrast, Al and Mn are the least reported elements. The lack of studies on trace elements in penguins from the coasts of Australia, South Africa, and Galapagos Islands is clearly noted. Most of the field studies of trace elements are concentrated in the Antarctic and subantarctic areas (>85%), specifically in the Antarctic Peninsula and South Shetland Islands; the rest of the studies are concentrated in the coasts of South America, Subtropical Front (Indian Ocean), and coasts of Australia. Further studies are needed in order to overcome the huge gap of data between Antarctica and other territories of more temperate zones where there are colonies of other species of penguins. Differences in trace element concentrations in the same species at different sites are evidenced in gentoo penguins, because they have a large distribution and a very plastic diet depending on site. Gentoo penguins at Crozet Islands have higher feather Hg concentrations (Carravieri et al. 2016) than those reported at Antarctic locations (Bargagli et al. 1998). Gentoo penguins at subantarctic areas have higher feather Hg concentrations than those reported at Antarctic locations (Table 1). Gentoo penguins at higher latitudes feed largely on krill (Carlini et al. 2009), whereas they prey mainly on fish at lower latitudes (Lescroël et al. 2004). Krill is a pelagic low-trophic prey very abundant in Antarctica with lower Hg burden compared to fish (Bargagli et al. 1998; Bustamante et al. 2003).

Due to the lack of data, the comparison of the levels of trace elements among different species and populations of penguins must be taken with caution. In general, the concentrations of trace elements are fragmented from the spatio-temporal point of view, which prevents for now conducting an analysis of tendencies. Hence, the implementation of monitoring programs that incorporate these variables is required.

4.3 Interspecific Differences

There are 18 species of penguins that inhabit the planet (García and Boersma 2013), but trace metals have been reported only in 11 species, evidencing the information gap in species such as *Eudyptes pachyrhynchus, Eudyptes sclateri, Eudyptes robustus, Eudyptes schlegeli, Spheniscus demersus, Spheniscus mendiculus,* and *Megadyptes antipodes.* The species with more elements reported are *P. papua,*

Table 10 Locations and number of studies performed on trace elements in different species of penguins worldwide

Region	Location	Ψ	Coordinates	Studies[a]	Metals reported	References
West Antarctica	Barton peninsula	1	62°14′S, 58°46′W	1	Pb, Hg	Yin et al. (2008)
	Potter cove	2	62°14′16″S, 58°39′59″W	1	As, cd, Pb, Hg, cu, Mn	Smichowski et al. (2006)
	Stranger point	3	62°15′32.00″S, 58°36′54.00″W	1	Cd, Pb, cu, Zn, Mn	Celis et al. (2015b)
	Arctowski	4	62° 9′36″S, 58°28′25″W	1	As, cd, Pb, Hg, cu, Zn	Celis et al. (2015a)
	King George Island	5	62°02′S, 58°21′W	4	Al, as, cd, Pb, Hg, cu, Zn, Mn	Brasso et al. (2014); Jerez et al. (2013a, b)
	Admiralty Bay	6	62° 4′52″S, 58°23′41″W	1	Hg, Zn	Santos et al. (2006)
	Narebski point	7	62°12′S, 58°45′W	1	As, cd, Pb, Hg, cu, Zn	Espejo et al. (2014)
	Fildes peninsula	8	62°12′S, 58°58′W	1	Pb, Hg	Yin et al. (2008)
	Yankee Harbor	9	62°31′60.00″S, 59°46′41.0″W	1	As, cd, Pb, cu, Zn	Espejo et al. (2014)
	Livingston Island	10	62°37′S 60°27′W	4	Al, as, cd, Pb, cu, Zn, Mn	Metcheva et al. (2006, 2010, 2011); Jerez et al. (2011)
	Cape Shirreff	11	62°28′S, 60°47′W	1	As, cd, Pb, cu, Zn	Espejo et al. (2014)
	Byers peninsula, Livingston is.	12	62°38′S, 61°05′W	1	Al, as, cd, Pb, cu, Zn, Mn	Barbosa et al. (2013)
	Hannah point, Livingston is.	13	62°39′16″S, 60°36′48″W	1	Al, as, cd, Pb, cu, Zn, Mn	Barbosa et al. (2013)
	Deception Island	14	62°56′27″S, 60°35′39″W	3	Al, as, cd, Pb, cu, Zn, Mn	Jerez et al. (2011); Jerez et al. (2013a, b)

(continued)

Table 10 (continued)

Region	Location	Ψ	Coordinates	Studies[a]	Metals reported	References
	O'Higgins Base	15	63°19′15″S, 57°53′55″W	3	Al, as, cd, Pb, Hg, cu, Zn, Mn	Celis et al. (2012); Espejo et al. (2014), Celis et al. (2015b)
	Kopaitic	16	63°18′59″S, 57°54′47″W	2	As, cd, Pb, Hg, cu, Zn	Celis et al. (2015a), Espejo et al. (2014)
	Mikkelsen Harbor	17	63°53′22″S, 60°47′3″W	1	As, cd, Pb, Hg, cu, Zn	Espejo et al. (2014)
	Hydrurga rocks	18	64° 8′40″S, 61°40′22″W	1	As, cd, Pb, cu, Zn	Espejo et al. (2014)
	Danco Island	19	64°43′53″S, 62°35′44″W	1	As, cd, Pb, cu, Zn	Espejo et al. (2014)
	Ronge Island	20	64°43′S, 62°41′W	1	Al, as, cd, Pb, cu, Zn, Mn	Jerez et al. (2011)
	Neko Harbor	21	64°50′S, 62°33′W	1	Cd, Pb, cu, Zn, Mn	Celis et al. (2015b)
	González Videla Base	22	64° 49′ 26″ S, 62° 51′ 26″ W	2	As, cd, Pb, Hg, cu, Zn	Celis et al. (2012); Espejo et al. (2014)
	Brown Station	23	64°53′43.2″S, 62°52′13.7″W	1	As, cd, Pb, cu, Zn	Espejo et al. (2014)
	Paradise Bay	24	64°53′S, 62°53′W	1	Al, as, cd, Pb, cu, Zn, Mn	Jerez et al. (2011)
	Yalour Island	25	65°14′2″S, 64°13′26″W	2	Al, as, cd, Pb, Hg, cu, Zn, Mn	Jerez et al. (2011); Celis et al. (2015a)
	Yelcho Station	26	64°52′33″S 63°35′02″W	1	As, cd, Pb, cu, Zn	Espejo et al. (2014)
	Doumer Island	27	64°51′S, 63°35′W	1	Cd, Pb, cu, Zn, Mn	Celis et al. (2015b)

	Avian Island	28	67°46′12″S, 68°53′40″W	2	Al, as, cd, Pb, Hg, cu, Zn, Mn	Celis et al. (2015a); Jerez et al. (2013a)
	Weddell Sea	29	77°S 49°W	1	Cd, cu	Steinhagen-Schneider (1986)
East Antarctica	Edmonson point	30	74°20′S 165°8′E	1	Cd, Pb, Hg	Ancora et al. (2002)
	Terra Nova Bay	31	74°47′30″S, 164°51′35″E	1	Hg	Bargagli et al. (1998)
	Adélie land	32	66°40′ S, 140°01′ E	1	Hg	Carravieri et al. (2016)
	Zhongshan Station	33	69°22′ S, 76°22′ E	1	Pb, Hg	Yin et al. (2008)
West coast of South America	Pan de Azúcar is., Chile	34	26° 9′S, 70°41′29″W	1	As, cd, Pb, Hg, cu, Zn	Celis et al. (2014)
	Chañaral is., Chile	35	29° 1′33″S, 71°34′5″W	1	As, cd, Pb, Hg, cu, Zn	Celis et al. (2014)
	Cachagua is., Chile	36	32°35′6″S, 71°27′24″W	1	As, cd, Pb, Hg, cu, Zn	Celis et al. (2014)
East Coast of South America	Rio Grande do Sul, Brazil	37	31°11′20″S, 50°51′47″W	1	Cd, Pb, Hg	Kehrig et al. (2015)
	Punta Tombo, Argentina	38	44° 2′17″S, 65°12′2″W	1	Hg	Frias et al. (2012)
Subantarctic zone	South Georgia is.	39	54° 16′ 53″ S, 36° 30′ 28″ W	1	Hg	Pedro et al. (2015)
	Crozet Islands	40	46°26′S, 51°45′E	2	Hg	Scheifler et al. (2005); Carravieri et al. (2016)
	Kerguelen Islands	41	49°21′S, 70°18′E	1	Hg	Carravieri et al. (2013)
Subtropical front, Indian Ocean	Amsterdam Island	42	37°50′ S, 77°31′ E	1	Hg	Carravieri et al. (2016)
Coast of Australia	Victoria	43	38°48′53″S, 146° 5′36″E	1	Al, as, cd, Pb, Hg, cu, Zn	Finger et al. (2015)

Ψ Number of the position pointed in Fig. 3
[a]Number of studies performed

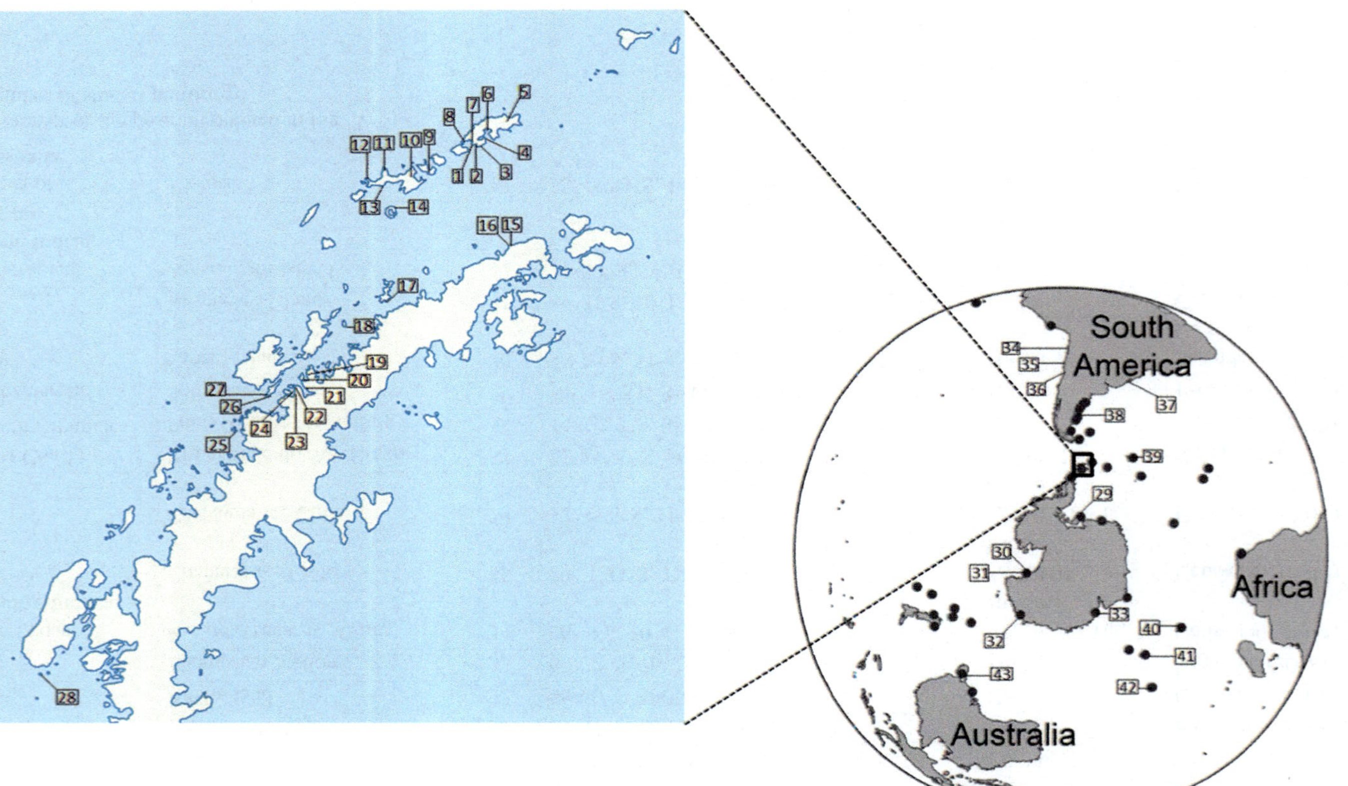

Fig. 3 Geographical distribution of trace metals reported in penguins. Each number is associated with the information given in Table 10

P. adeliae, and *P. antarctica*. On the other hand, the least studied species are *E. chrysocome* and *E. chrysolophus*. It is necessary to state that the distribution of trace elements by species from different studies, species, and individuals presents serious limitations because of temporal variation, spatial variation, diet, individual specialization, physiological condition, and sex.

In Adélie, gentoo, and chinstrap penguins, concentrations of the essential trace elements Cu, Mn, and Zn in all biotic matrices exhibited less inter-species variation than the non-essential Al, As, Cd, Hg, and Pb, expressed through the coefficient of variation. These results are in agreement with similar findings of other investigations in seabirds (Honda et al. 1990; Lock et al. 1992). Penguins, by virtue of their members exhibiting a wide range of trace element burdens, along with variation in diet, varying moult strategies and variation in their average life spans may explain inter-species pattern of metal accumulation, storage, and elimination (Thompson 1990; García and Boersma 2013). This is an issue that needs to be investigated more.

Diet is one of the most important factors that explain differences in trace element concentrations among the species. Penguins are useful bioindicators of Hg contamination in their food webs (Carravieri et al. 2016). Feather Hg concentrations in *Eudyptes* and *Pygoscelis* penguins are lower than *Aptenodytes* penguins, because they feed at lower trophic levels (Carravieri et al. 2013). One study showed that the concentrations of Zn, Al, and Mn in feathers were significantly higher in gentoo than in chinstrap penguins, which could be explained by the different diet and feeding habits of these species (Metcheva et al. 2006). During the Antarctic summer, when the breeding season takes place, gentoo penguins feed inshore, eating mainly crustaceans (68%) and fish (32%), even though foraging areas may also be included in their diet (Croxall et al. 1997). Also chinstrap penguins feed almost exclusively on krill, but can feed beyond the continental shelf areas (Espejo et al. 2014). Concentration of trace elements can differ among colonies of the same species that live far from each other owing to diet and the presence of chemicals in waters (Jerez et al. 2011). Similarly, Yin et al. (2008) mention that the difference in Cu levels among seabirds might be related to different food resources for the species. The trophic level of the species which is given by diet can be determined by means of stable isotopes of nitrogen, a method infrequently used in studies of trace elements in penguins (Brasso and Polito 2013; Brasso et al. 2014; Carravieri et al. 2016).

5 Summary and Conclusions

In the environment, trace elements are persistent and come from both natural cycling in the biosphere and anthropogenic activities (Nordberg and Nordberg 2016). For this reason there is a concern about the possible negative effects these contaminants may have on animals and marine ecosystems (Szopińska et al. 2016). Birds are excellent indicators of the degree of pollution in the environment, because they rapidly express the biological impacts of the contaminants that can even be extrapolated to humans (Ochoa-Acuña et al. 2002; Cifuentes et al. 2003; Zhang and Ma 2011), a remarkable issue considering humans are the most sensitive species to the toxic effects of some trace elements (Byrns and Penning 2011). The human population will increase and also increase marine-derived protein consumption. Most penguins include fish in their diets, such as sprats, myctophid fish, anchovies, silversides, jack mackerel, and common hake, among others (García and Boersma 2013). Many fishes are also consumed by humans, thus these birds might be used as a bioindicator for human health and as exposure assessment. Most penguins are on the upper side of the trophic chain and they depend on few species for food. Consequently, the effects on a particular species might loom as a serious threat to penguins.

Investigations of trace elements in penguins report mostly the levels of Al, As, Cd, Cu, Hg, Mn, Pb, and Zn. The most reported metal is Pb, whereas Al is the least reported. Other metals such as Co, Cr, Fe, or Ni have been poorly studied (Jerez et al. 2013a; Szopińska et al. 2016). The oldest data dates back to the 1950s and it was aimed at determining the Hg levels in feathers of King penguins (*Aptenodytes patagonicus)* from Crozet Islands, South East of Indian Ocean (Carravieri et al. 2016).

There are 18 species of penguins around the world and trace elements have been reported in 11 of them (*P. papua, P. antarctica, P. adeliae, Aptenodytes forsteri, A. patagonicus, Spheniscus magellanicus, S. humboldti, Eudyptes chrysolophus, E. chrysocome, E. minor,* and *E. moseleyi*). Most studies of concentrations of trace elements in penguins have been focused on the genus *Pygoscelis*, mainly on gentoo penguins, followed by Adélie penguins, and finally chinstrap penguins. Other penguin species such as *E. pachyrhynchus, E. sclateri, E. robustus, E. schlegeli, S. demersus, S. mendiculus,* and *Megadyptes antipodes* have not received any attention.

The most studied penguin biological matrices are feathers and then excreta, followed by the liver, kidneys, bones, muscles, and stomach contents. On the other hand, studies carried out to measure trace elements in blood and internal organs such as heart, testicles, spleen, or brains of penguins are scarce (Bargagli et al. 1998; Finger et al. 2015; Metcheva et al. 2010; Metcheva et al. 2011). The species which display the highest concentration of most trace elements are the gentoo penguin (33%), followed by the Adélie penguin (31%), the chinstrap penguin (19%), the Humboldt penguin (7%), the Magellanic penguin (6%), and the Emperor penguin (4%).

The maximum concentrations (μg/g, dw) of Al (2595) have been found in stomach contents of gentoo penguins from King George Island, and Cd (351.8) in the liver of Adélie penguins from Antarctic Peninsula. The highest levels of As (7.9) and Pb (12.8) were found in excreta of Humboldt penguins from the Central Coast of Chile. Maximum concentrations of Hg (6.6) and Cu (585.8) have been reported in excreta of gentoo penguin and Adélie penguin, respectively, both from the Antarctic Peninsula, whereas the maximum Zn levels (487.1) was found in excreta of Humboldt penguins of Northern Chile. Finally, excepting excreta and stomach contents, maximum levels of Mn (18.35) are in the bones of gentoo penguins from Byers Peninsula (South Shetland Islands). The large variation in trace element concentrations detected in different biotic matrices of penguins in Antarctica might be explained because in this continent many pristine places coexist with locations having major human presence, a situation which rarely occurs in others areas of the world. Additionally, several other factors can force variation in trace element concentrations in penguin tissues such as feeding ecology, physiological state, species, age class, molting patterns, among others.

In general, Hg, Pb, and Cd concentrations in penguins are lower than those reported in other seabirds from the Northern Hemisphere, whereas the concentrations of Al and As are otherwise. The concentrations of Cu, Mn, and Zn tend to be within the range reported in the marine birds of the Northern Hemisphere, suggesting that those elements are regulated in seabirds (Gibbs 1995). The highest levels of Cu and Cd correspond to penguins that live in Antarctica, which might be related to the high levels of these metals detected in the Antarctic krill (Nygard et al. 2001). On the other hand, it has been observed that in the Antarctic Peninsula there is a natural enrichment of Cd, As, and Al in the trophic chains, due to the local volcanism (Deheyn et al. 2005). Nevertheless, comparisons could be influenced by the differences in the diet composition of each of the species (Jerez et al. 2011).

Studies on the effects of trace elements on penguins are scarce. For that reason, the comparison of data reported in penguins with those obtained from studies performed on birds at other locations and ecologically different to penguins was unavoidable. Hence, any comparison to toxic thresholds of trace elements in terrestrial birds should be taken with extreme caution, because seabirds appear to be more resistant to toxic effect of most pollutants than are mammals or terrestrial birds (Beyer et al. 1996). In general, the concentrations of trace elements in the different organs of penguins are below the toxicity thresholds with negative biological consequences for seabirds. Some colonies of Humboldt penguins located in areas with human presence on the coast of Chile might present some pathological problems due to As, Cu, and Pb (Celis et al. 2014). Some negative effects in the liver and kidneys of gentoo penguins from the Antarctic Peninsula could be linked to local Pb contamination (Celis et al. 2012, Jerez et al. 2013a). Levels of Zn in livers of some colonies of gentoo and chinstrap penguins from King George Island (Jerez et al. 2013a) exceeded in 19% and 65% the threshold value of physiological importance for seabirds, respectively (Honda et al. 1990). It seems to be related to areas of greatest human activities in Antarctica, which are concentrated precisely on King George Island (Bargagli 2008; Tin et al. 2009). Levels of Cd in livers of some colonies of gentoo, Adélie, chinstrap, and Emperor penguins that inhabit the Antarctic Peninsula area, and Magellanic penguins from southern Brazil, which together represent almost 48% of the reported colonies might be associated with physiological and ecological problems (>3 μg/g, Scheuhammer 1987). Cadmium concentrations found in kidneys of Adélie, chinstrap, and Emperor penguin from some locations of the Antarctic Peninsula (270.2–351.8 μg/g, Table 4), such as Avian Island, Deception Island, and Weddell Sea, indicate that these birds have suffered some degree of chronic exposure to this metal (Furness 1996). Further studies that correlate the levels of trace metals found in non-invasive samples with biological effects on penguins are required.

Most studies of concentrations of trace elements in penguins have been carried out on the Antarctica and subantarctic islands, thus evidencing a lack of data from other areas where penguins live also, such as Australia, South Africa, South America, and Galapagos Islands. It is surprising to find studies mainly in Antarctica, since researchers require an adequate implementation and a firm determination to work under extreme climatic conditions. Perhaps the urge to travel to remote and poorly explored regions is more important than the simple desire of performing research in more populated places with more temperate climates where the species of threatened penguins could be more exposed to contaminants by being in areas with major human presence.

The trophic transfer coefficient, calculated from the levels of metals available in gentoo, chinstrap, and Adélie penguins, suggests a possible biomagnification of Cd and Zn. Due to the fact that scientists have always believed that metals, except Hg, are not biomagnified, this issue needs to be studied more in different environments inhabited by penguins.

Most studies of penguins have focused on measuring the levels of exposure in different biotic matrices. The concentration of metals in tissues and organs of penguins may have a great toxicological importance. In humans, diseases related to deficiency of essential trace elements are well known (Nordberg and Nordberg 2016). Further studies with biomarkers are needed in order to evaluate the actual risks associated with the levels of these contaminants in polar environments with low ecological diversity, which can increase diseases with consequences for the health of penguin populations (Boersma 2008).

Little is known about the interaction of metals that might activate certain detoxification mechanisms of the organism of penguins. It is suspected that Se could play an important role in the detoxification processes of Hg. The study with species in captivity could be a good alternative to evaluate the physiological mechanisms of these species at a given concentration of metals, under a controlled environment (Falkowska et al. 2013).

In the short term, studies of trace elements in penguins should take into account the following aspects:

- Incorporation of other metals such as Co, Ni, or Cr and their possible effects in the organisms of different species of penguins in order to perform more accurate risk assessments.
- Further toxico-kinetics studies of trace element levels in penguins, including other tissues and organs, are needed to better understand the overall toxicity in seabirds.
- Information on metals of the following species is crucial: *Eudyptes pachyrhynchus, Eudyptes moseleyi, Eudyptes sclateri, Eudyptes robustus, Eudyptes schlegeli, Spheniscus demersus, Spheniscus mendiculus,* and *Megadyptes antipodes.*
- Correlation between the levels of metals in different biological matrices with their effects on different species and geographic locations is required.

- Interspecific variation of metals should be addressed more in depth, with isotopes of nitrogen being a good tool to understand differences among species.
- The implementation of monitoring programs that incorporate spatial-temporal data is required for conducting an analysis of tendencies.
- It is crucial to implement uniform monitoring protocols to help unify the data and make it more comparable.

Acknowledgments Winfred E. Espejo is a graduate student at the Universidad de Concepción, Chile, who is sponsored by the CONICYT-Chile to pursue PhD research. This study was financially supported by the project INACH RG 09-14 (J. Celis), INACH T31-11 (G. Chiang), and FONDAP CRHIAM 15 13 0015 (R. Barra). Thanks also are given to project 216.153.025-1.0 of the Research Division of the Universidad de Concepción. Many thanks are also given to Dr. Evelyn Habit, Liseth Chaura, and peer reviewers for their useful suggestions. Finally, the authors also thank Diane Haughney for the English revision.

References

Agusa T, Matsumoto T, Ikemoto T, Anan Y, Kubota R, Yasunaga G, Kunito T, Tanabe S, Ogi H, Shibata Y (2005) Body distribution of trace elements in black-tailed gulls from Rishiri island, Japan: age-dependent accumulation and transfer to feathers and eggs. Environ Toxicol Chem 24:2107–2120

Anan Y, Kunito T, Watanabe I, Sakai H, Tanabe S (2001) Trace element accumulation in hawksbill turtle (*Eretmochelys imbricata*) and green turtle (*Chelonia mydas*) from Yaeyama Islands, Japan. Environ Toxicol Chem 20:2802–2814

Ancora S, Volpi V, Olmastroni S, Focardi S, Leonzio C (2002) Assumption and elimination of trace elements in Adélie penguins from Antarctica: a preliminary study. Mar Environ Res 54:341–344

Arcos JM, Ruiz X, Bearhop S, Furness RW (2002) Mercury levels in seabirds and their fish prey at the Ebro Delta (NW Mediterranean): the role of trawler discards as a source of contamination. Mar Ecol Prog Ser 232:281–290

ATSDR (2004) Toxicological profile for copper. Agency for Toxic Substances and Disease Registry US Department of Health and Human Services, Public Health Service, Atlanta, GA. http://www.atsdr.cdc.gov/ToxProfiles/tp132.pdf. Accessed 19 May 2016

ATSDR (2007) Toxicological profile for arsenic. Agency for Toxic Substances and Disease Registry US Department of Health and Human Services, Public Health Service, Atlanta, GA. http://www.atsdr.cdc.gov/toxprofiles/tp2.pdf. Accessed 19 May 2016

ATSDR (2008) Toxicological profile for manganese. Agency for Toxic Substances and Disease Registry US Department of Health and Human Services, Public Health Service, Atlanta, GA. http://www.atsdr.cdc.gov/toxprofiles/tp.asp?id=102&tid=23. Accessed 19 May 2016

Ballatori N (2002) Transport of toxic metals by molecular mimicry. Environ Health Perspect 110:689–694

Barbieri E, de Andrade PE, Filippini A, Souza dos Santos I, Borges CA (2010) Assessment of trace metal concentration in feathers of seabird (*Larus dominicanus*) sampled in the Florianópolis, SC, Brazilian coast. Environ Monit Assess 169:631–638

Barbosa A, De Mas E, Benzal J, Diaz J, Motas M, Jerez S, Pertierra L, Benayas J, Justel A, Lauzurica P, Garcia-Peña F, Serrano T (2013) Pollution and physiological variability in gentoo penguins at two rookeries with different levels of human visitation. Antarct Sci 25:329–338

Bargagli R (2001) Trace metals in Antarctic organisms and the development of circumpolar biomonitoring networks. Rev Environ Contam Toxicol 171:53–110

Bargagli R (2008) Environmental contamination in Antarctic ecosystems. Sci Total Environ 400:212–226

Bargagli R, Monaci F, Sánchez-Hernández J, Cateni D (1998) Biomagnification of mercury in an Antarctic marine coastal food web. Mar Ecol Prog Ser 169:65–76

Barjaktarovic L, Elliott JE, Scheuhammer AM (2002) Metal and metallothionein concentrations in scoter (*Melanitta* spp.) from the Pacific northwest of Canada, 1989–1994. Arch Environ Contam Toxicol 43:486–491

Becker PH, González-Solís J, Behrends B, Croxall J (2002) Feather mercury levels in seabirds at South Georgia: influence of trophic position, sex and age. Mar Ecol Prog Ser 243:261–269

Beyer WN, Heinz GH, Redmon-Norwood AW (1996) Environmental contaminants in wildlife: interpreting tissue concentrations. Lewis, Boca Raton, FL, p 512

Beyer NW, Spalding M, Morrison D (1997) Mercury concentrations in feathers of wading birds from Florida. Ambio 26:97–100

Beyer WN, Franson JC, Locke LN, Stroud RK, Sileo L (1998) Retrospective study of the diagnostic criteria in a lead-poisoning survey of waterfowl. Arch Environ Contam Toxicol 35:506–512

Boersma PD (2008) Penguins as marine sentinels. Bioscience 58:597–607

Brasso RL, Polito MJ (2013) Trophic calculations reveal the mechanism of population-level variation in mercury concentrations between marine ecosystems: case studies of two polar seabirds. Mar Pollut Bull 75:244–249

Brasso RL, Polito MJ, Emslie SD (2014) Multi-tissue analyses reveal limited inter-annual and seasonal variation in mercury exposure in an Antarctic penguin community. Ecotoxicology 23:1494–1504

Braune BM, Gaston AJ, Hobson KA, Gilchrist HG, Mallory ML (2014) Changes in food web structure alter trends of mercury uptake at two seabird colonies in the Canadian Arctic. Environ Sci Technol 48:13246–13252

Bryan GW, Langston WJ (1992) Bioavailability, accumulation and effects of heavy metals in sediments with special reference to United Kingdom estuaries: a review. Environ Pollut 76:89–131

Burger J (2008) Assessment and management of risk to wildlife from cadmium. Sci Total Environ 389:37–45

Burger J, Gochfeld M (1997) Risk, mercury levels, and birds: relating adverse laboratory effects to field biomonitoring. Environ Res 75:160–172

Burger J, Gochfeld M (2000a) Effects of lead on birds (*Laridae*): a review of laboratory and field studies. J Toxicol Environ Health B Crit Rev 3:59–78

Burger J, Gochfeld M (2000b) Metal levels in feathers of 12 species of seabirds from midway atoll in the northern Pacific Ocean. Sci Total Environ 257:37–52

Burger J, Gochfeld M, Sullivan K, Irons D, McKnight A (2008) Arsenic, cadmium, chromium, lead, manganese, mercury, and selenium in feathers of black-legged kittiwake (*Rissa tridactyla*) and black oystercatcher (*Haematopus bachmani*) from Prince William sound, Alaska. Sci Total Environ 398:20–25

Burger J, Tsipoura N, Newhouse M, Jeitner C, Gochfeld M, Mizrahi D (2011) Lead, mercury, cadmium, chromium, and arsenic levels in eggs, feathers, and tissues of Canada geese of the New Jersey meadowlands. Environ Res 111:775–784

Bustamante P, Bocher P, Cherel Y, Miramand P, Caurant F (2003) Distribution of trace elements in the tissues of benthic and pelagic fish from the Kerguelen Islands. Sci Total Environ 313:25–39

Byrns MC, Penning TM (2011) Environmental toxicology. Carcinogens and heavy metals. In: Brunton L, Chabner B, Knollman B (eds) The pharmacological basis of therapeutics. McGraw Hill, New York, pp 1853–1878

Calle P, Alvarado O, Monserrate L, Cevallos JM, Calle N, Alava JJ (2015) Mercury accumulation in sediments and seabird feathers from the Antarctic peninsula. Mar Pollut Bull 91:410–417

Campbell LM, Norstrom RJ, Hobson KA, Muir D, Backus S, Fisk AT (2005) Mercury and other trace elements in a pelagic Arctic marine food web (Northwater polynya, Baffin Bay). Sci Total Environ 351-352:247–263

Carlini AR, Coria NR, Santos MM, Negrete J, Juares MA, Daneri GA (2009) Responses of *Pygoscelis adeliae* and *P. papua* populations to environmental changes at Isla 25 de Mayo (king George Island). Polar Biol 32:1427–1433

Carravieri A, Bustamante P, Churlaud C, Cherel Y (2013) Penguins as bioindicators of mercury contamination in the Southern Ocean: birds from the Kerguelen Islands as a case study. Sci Total Environ 454-455:141–148

Carravieri A, Bustamante P, Churlaud C, Fromant A, Cherel Y (2014) Moulting patterns drive within-individual variations of stable isotopes and mercury in seabird body feathers: implications for monitoring of the marine environment. Mar Biol 161:963–968

Carravieri A, Cherel Y, Jaeger A, Churlaud C (2016) Penguins as bioindicators of mercury contamination in the southern Indian Ocean: geographical and temporal tends. Environ Pollut 213:195–205

Casini S, Fossi M, Gavilan J, Barra R, Parra O, Leonzio C, Focardi S (2001) Porphyrin levels in excreta of seabirds of the Chilean coasts as nondestructive biomarker of exposure to environmental pollutants. Arch Environ Contam Toxicol 4:65–72

Casini S, Fossi M, Leonzio C, Renzoni A (2003) Review: porphyrins as biomarkers for hazard assessment of bird populations: destructive and non-destructive use. Ecotoxicology 12:297–305

Celis J, Jara S, González-Acuña D, Barra R, Espejo W (2012) A preliminary study of trace metals and porphyrins in excreta of gentoo penguins (*Pygoscelis papua*) at two locations of the Antarctic peninsula. Arch Med Vet 44:311–316

Celis JE, Espejo W, González-Acuña D, Jara S, Barra R (2014) Assessment of trace metals and porphyrins in excreta of Humboldt penguins (*Spheniscus humboldti*) in different locations of the northern coast of Chile. Environ Monit Assess 186:1815–1824

Celis JE, Espejo W, Barra R, Gonzalez-Acuña D, Gonzalez F, Jara S (2015a) Assessment of trace metals in droppings of Adélie penguins (*Pygoscelis adeliae*) from different locations of the Antarctic peninsula area. Adv Polar Sci 26:1–7

Celis JE, Barra R, Espejo W, González-Acuña D, Jara S (2015b) Trace element concentrations in biotic matrices of gentoo penguins (*Pygoscelis papua*) and coastal soils from different locations of the Antarctic peninsula. Water Air Soil Pollut 226:1–12

Chen CY, Serrell N, Evers DC, Fleishman BJ, Lambert KF, Weiss J, Mason RP, Bank MS (2008) Meeting report: methylmercury in marine ecosystems—from sources to seafood consumers. Environ Health Perspect 116:1706–1712

Cifuentes JM, Becker PH, Sommer U, Pacheco P, Schlatter R (2003) Seabird eggs as bioindicators of chemical contamination in Chile. Environ Pollut 126:123–137

Clarke KR, Somerfield PJ, Chapman MG (2006) On resemblance measures for ecological studies, including taxonomic dissimilarities and a zero-adjusted bray–Curtis coefficient for denuded assemblages. J Exp Mar Biol Ecol 330:55–80

Cooper J, Crawford RJM, De Villiers M, Dyer BM, Hofmeyr GJG, Jonker A (2009) Disease outbreaks among penguins at sub-Antarctic Marion Island: a conservation concern. Mar Ornithol 37:193–196

Croxall JP, Prince PA, Reid K (1997) Dietary segregation of krill eating South Georgia seabirds. J Zool 242:531–556

Custer T, Custer C, Eichhorst B, Warburton D (2007) Selenium and metal concentrations in waterbird eggs and chicks at Agassiz National Wildlife Refuge, Minnesota. Arch Environ Contam Toxicol 53:103–109

Daso AP, Okonkwo JO, Jansen R, Brandao JD, Kotzé A (2015) Mercury concentrations in eggshells of the southern ground-hornbill (*Bucorvus leadbeateri*) and Wattled crane (*Bugeranus carunculatus*) in South Africa. Ecotoxicol Environ Saf 114:61–66

Dauwe T, Bervoets L, Blust R, Pinxten R, Eens M (2000) Can excrement and feathers of nestling songbirds be used as biomonitors for heavy metal pollution? Arch Environ Contam Toxicol 39:541–546

Dauwe T, Janssens E, Bervoets L, Blust R, Eens M (2005) Heavy-metal concentrations in female laying great tits (*Parus major*) and their clutches. Arch Environ Contam Toxicol 49:249–256

De Matos R (2008) Calcium metabolism in birds. Vet Clin Exot Anim 11:59–82

De Matteis F, Lim CK (1994) Porphyrins as nondestructive indicators of exposure to environmental pollutants. In: Fossi MC, Leoncio C (eds) Nondestructive biomarkers in vertebrates. Lewis, Boca Raton, FL, pp 93–128

De Moreno JEA, Gerpe MS, Moreno VJ, Vodopivez C (1997) Heavy metals in Antarctic organisms. Polar Biol 17:131–140

Debacker V, Jauniaux T, Coignoul F, Bouquegneau JM (2000) Heavy metals contamination and body condition of wintering guillemots (Uria Aalge) at the Belgian coast from 1993 to 1998. Environ Res 84(3):310–317

Deheyn DD, Gendreau P, Baldwin RJ, Latz MI (2005) Evidence for enhanced bioavailability of trace elements in the marine ecosystem of Deception Island, a volcano in Antarctica. Mar Environ Res 60:1–33

Dehn L, Follmann E, Thomas D, Sheffield G, Rosa C, Duffy L, O'Hara T (2006) Trophic relationships in an Arctic food web and implications for trace metal transfer. Sci Total Environ 362:103–123

Del Hoyo J, Elliott A, Sargatal J (1992) Handbook of the birds of the world. Volume 1: ostrich to ducks. Lynx Edicions. Barcelona, Spain, p 696

Dos Santos IR, Silva-Filho EV, Schaefer CE, Albuquerque-Filho MR, Campos LS (2005) Heavy metal contamination in coastal sediments and soils near the Brazilian Antarctic Station, king George Island. Mar Pollut Bull 50:185–194

Eisler R (1985) Selenium hazards to fish, wildlife and invertebrates. A synoptic review. U.S. Fish and Wildlife Service, Washington DC. https://nrm.dfg.ca.gov/FileHandler.ashx?DocumentID=7297. Accessed 31 May 2016

Eisler R (1987) Mercury hazards to fish, wildlife, and invertebrates: a synoptic review. U.S. Fish and WildlifeService, Washington DC. https://www.pwrc.usgs.gov/eisler/CHR_10_Mercury.pdf

Eisler R (1988) Lead hazards to fish, wildlife, and invertebrates. A synoptic review. U.S. Fish and Wildlife Service, Washington DC. https://www.pwrc.usgs.gov/eisler/CHR_14_Lead.pdf. Accessed 31 May 2016

Eisler R (1993) Zinc hazards to fish, wildlife and invertebrates. A synoptic review. U.S. Fish and Wildlife Service, Washington DC. https://www.pwrc.usgs.gov/eisler/CHR_26_Zinc.pdf. Accessed 31 May 2016

Eisler R (1998) Copper hazards to fish, wildlife, and invertebrates. A synoptic review. U.S. Fish and Wildlife Service, Washington DC. https://www.pwrc.usgs.gov/eisler/CHR_33_Copper.pdf. Accessed 31 May 2016

Elliot JE, Scheuhammer AM, Leighton FA, Pearce PA (1992) Heavy metal and metallothionein concentrations in Atlantic Canadian seabirds. Arch Environ Contam Toxicol 22:63–73

Erikson KM, Aschner M (2003) Manganese neurotoxicity and glutamate-GABA interaction. Neurochem Int 43:475–480

Espejo W, Celis J, González-Acuña D, Jara S, Barra R (2014) Concentration of trace metals in excrements of two species of penguins from different locations of the Antarctic peninsula. Polar Biol 37:675–683

Evers DC, Savoy LJ, DeSorbo CR, Yates DE, Hanson W, Taylor KM, Siegel LS, Cooley JH Jr, Bank MS, Major A, Munney K, Mower BF, Vogel HS, Schoch N, Pokras M, Goodale MW, Fair J (2008) Adverse effects from environmental mercury loads on breeding common loons. Ecotoxicology 17:69–81

Falkowska L, Reindl AR, Szumilo E, Kwaśniak J, Staniszewska M, Bełdowska M, Lewandowska A, Krause I (2013) Mercury and chlorinated pesticides on the highest level of the food web as exemplified by herring from the southern Baltic and African penguins from the zoo. Water Air Soil Pollut 224:1549

Fimreite N (1974) Mercury contamination of aquatic birds in northwestern Ontario. J Wildl Manag 38:120–131

Finger A, Lavers JL, Dann P, Nugegoda D, Orbell JD, Robertson B, Scarpaci C (2015) The little penguin (*Eudyptula minor*) as an indicator of coastal trace metal pollution. Environ Pollut 205:365–377

Fitzgerald WF, Lamborg CH, Hammerschmidt CR (2007) Marine biogeochemical cycling of mercury. Chem Rev 107:641–662

Franson JC (1996) Interpretation of tissue lead residues in birds other than waterfowl. In: Beyer WN, Heinz GH, Redmon-Norwood AW (eds) Environmental contaminants in wildlife. Interpreting tissue concentrations. Lewis, Boca Raton, FL, pp 264–279

Franson JC, Hollmén TE, Flint PL, Grand JB, Lanctot RB (2003) Contaminants in molting long-tailed ducks and nesting common eiders in the Beaufort Sea. Mar Pollut Bull 45:504–513

Franson JC, Hoffman DJ, Wells-Berlin A, Perry MC, Shearn-Bochsler V, Finley DL, Flint PL, Hollmén T (2007) Effects of dietary selenium on tissue concentrations, pathology, oxidative stress, and immune function in common eiders (*Somateria mollissima*). J Toxicol Environ Health A 70:861–874

Frias JE, Gil MN, Esteves JL, Borboroglu PG, Kane OJ, Smith JR, Boersma PD (2012) Mercury levels in feathers of Magellanic penguins. Mar Pollut Bull 64:1265–1269

Furness RW (1996) Cadmium in birds. In: Beyer WN, Heinz GH, Redmon-Norwood AW (eds) Environmental contaminants in wildlife. Interpreting tissue concentrations. Lewis, Boca Raton, FL, pp 389–404

Furness RW, Muirhead SJ, Woodburn M (1986) Using bird feathers to measure mercury in the environment: relationships between mercury content and molt. Mar Pollut Bull 17:27–30

García P, Boersma PD (2013) Penguins: natural history and conservation. University of Washington Press, Seattle & London, p 328

Gasparik J, Vladarova D, Capcarova M, Smehyl P, Slamecka J, Garaj P, Stawarz R, Massanyi P (2010) Concentration of lead, cadmium, mercury and arsenic in leg skeletal muscles of three species of wild birds. J Environ Sci Health A 45:818–823

Gibbs PJ (1995) Heavy metal and organochlorine concentrations in tissues of the little penguin *Eudyptula minor*. In: Dann P, Norman I, Reilly P (eds) The penguins. Surrey Beatty & Sons, Australia, pp 393–419

Gilani SH, Alibhai Y (1990) Teratogenicity of metals to chick embryos. Toxicol Environ Health 30:23–31

Gochfeld M (1997) Spatial patterns in a bioindicator: heavy metal and selenium concentration in eggs of herring gulls (*Larus argentatus*) in the New York bight. Arch Environ Contam Toxicol 33:63–70

Gochfeld M, Belant JL, Shukla T, Benson T, Burger J (1996) Heavy metals in laughing gulls: gender, age and tissue differences. Environ Toxicol Chem 15:2275–2283

Goutner V, Furness R, Papakonstantinou K (2000) Mercury in feathers of Audouin's Gull (*Larus audouinii*) chicks from northeastern Mediterranean colonies. Arch Environ Contam Toxicol 39:200–204

Goutte A, Barbraud C, Herzke D, Bustamante P, Angelier F, Tartu S, Clément-Chastel C, Moe B, Bech C, Gabrielsen GW, Bustnes JO, Chastel O (2015) Survival rate and breeding outputs in a high Arctic seabird exposed to legacy persistent organic pollutants and mercury. Environ Pollut 200:1–9

Goyer RA (1997) Toxic and essential metal interactions. Annu Rev Nutr 17:37–50

Hoffman DJ (2002) Role of selenium toxicity and oxidative stress in aquatic birds. Aquat Toxicol 57:11–26

Honda K, Yamamoto Y, Hidaka H, Tatsukawa R (1986) Heavy metal accumulation in Adélie penguin, *Pygoscelis adeliae*, and their variations with the reproductive process. Mem Natl Inst Polar Res 40:443–453

Honda K, Marcovecchio JE, Kan S, Tatsukawa R, Ogi H (1990) Metal concentrations in pelagic seabirds from the North Pacific Ocean. Arch Environ Contam Toxicol 19:704–711

Iavicoli I, Fontana L, Bergamaschi A (2009) The effects of metals as endocrine disruptors. J Toxicol Environ Health B 12:206–223

Ikemoto T, Kunito T, Tanaka H, Baba N, Miyazaki N, Tanabe S (2004) Detoxification mechanism of heavy metals in marine mammals and seabirds: interaction of selenium with mercury, silver, copper, zinc, and cadmium in liver. Arch Environ Contam Toxicol 47:402–413

Jerez S, Motas M, Palacios MJ, Valera F, Cuervo JJ, Barbosa A (2011) Concentration of trace elements in feathers of three Antarctic penguins: geographical and interspecific differences. Environ Pollut 159:2412–2419

Jerez S, Motas M, Benzal J, Diaz J, Vidal V, D'Amico V, Barbosa A (2013a) Distribution of metals and trace elements in adult and juvenile penguins from the Antarctic peninsula area. Environ Sci Pollut R 20:3300–3311

Jerez S, Motas M, Benzal J, Diaz J, Barbosa A (2013b) Monitoring trace elements in Antarctic penguin chicks from south Shetland Islands, Antarctica. Mar Pollut Bull 69:67–75

Jin S, Seo S, Shin Y, Bing K, Kang T, Paek W, Lee D (2012) Heavy metal accumulations of 4 species of Anseriformes in Korea. J Korean Nat 5:345–349

Kaur N, Dhanju CK (2013) Heavy metals concentration in excreta of free living wild birds as indicator of environmental contamination. Bioscan 8:1089–1093

Kehrig HA, Hauser-Davis RA, Seixas TG, Fillmann G (2015) Trace-elements, methylmercury and metallothionein levels in Magellanic penguin (*Spheniscus magellanicus*) found stranded on the southern Brazilian coast. Mar Pollut Bull 96:450–455

Khan A, Hussain H, Sattar A, Khan M, Abbas R (2014) Toxico-pathological aspects of arsenic in birds and mammals: a review. Int J Agric Biol 16:1213–1224

Kim J, Koo T (2007) Heavy metal concentrations in diet and livers of black-crowned night heron *Nycticorax nycticorax* and grey heron *Ardea cinerea* chicks from Pyeongtaek, Korea. Ecotoxicology 16:411–416

Kim J, Oh J (2014a) Trace element concentrations in eggshells and egg contents of black-tailed gull (*Larus crassirostris*) from Korea. Ecotoxicology 23:1147–1152

Kim J, Oh J (2014b) Relationships of metals between feathers and diets of black-tailed gull (*Larus crassirostris*) chicks. Bull Environ Contam Toxicol 92:265–269

Kim J, Oh J (2014c) Heavy metal concentrations in black-tailed gull (*Larus crassirostris*) chicks, Korea. Chemosphere 112:370–376

Kim EY, Murakami T, Saeki K, Tatsukawa R (1996) Mercury levels and its chemical form in tissues and organs of seabirds. Arch Environ Contam Toxicol 30:259–266

Kim EY, Goto R, Tanabe S, Tanaka H, Tatsukawa R (1998) Distribution of 14 elements in tissues and organs of oceanic seabirds. Arch Environ Contam Toxicol 35:638–645

Kim M, Park K, Park J, Kwak I (2013) Heavy metal contamination and metallothionein mRNA in blood and feathers of black-tailed gulls (*Larus crassirostris*) from South Korea. Environ Monit Assess 185:2221–2230

Kler T, Vashishat N, Kumar M (2014) Heavy metals concentration in excreta of avian species from Ludhiana district. Int J Adv Res 2:873–879

Lalancette A, Morin Y, Measures L, Fournier M (2003) Contrasting changes of sensitivity by lymphocytes and neutrophils to mercury in developing grey seals. Dev Comp Immunol 27:735–747

Larison JR, Likens GE, Fitzpatrick JW, Crock JG (2000) Cadmium toxicity among wildlife in the Colorado Rocky Mountains. Nature 406:181–183

Lavoie R, Jardine T, Chumchal M, Kidd K, Campbell L (2013) Biomagnification of mercury in aquatic food webs: a worldwide meta-analysis. Environ Sci Technol 47:13385–13394

Lebedeva NV (1997) Accumulation of heavy metals by birds in the southwest of Russia. Russ J Ecol 28:41–46

Lee DP (1996) Relationship of heavy metal level in birds. Bull Kor Inst Orni 5:59–67

Lemley AD (1993) Guidelines for evaluating selenium data form aquatic monitoring and assessment studies. Environ Monit Assess 28:83–100

Lescroël A, Ridoux V, Bost CA (2004) Spatial and temporal variation in the diet of the gentoo penguin (*Pygoscelis papua*) at Kerguelen Islands. Polar Biol 27:206–216

Lim CK (1991) Porphyrins. In: Hanai T (ed) Liquid chromatography in biomedical analysis. Elsevier, Amsterdam, pp 209–229

Lock JW, Thompson DR, Furness RW (1992) Metal concentrations in seabirds of the New Zealand region. Environ Pollut 75:289–300

Lucia M, André JM, Gontier K, Diot N, Veiga J, Davail S (2010) Trace element concentrations (mercury, cadmium, copper, zinc, lead, aluminum, nickel, arsenic, and selenium) in some aquatic birds of the Southwest Atlantic Coast of France. Arch Environ Contam Toxicol 58:844–853

Majer A, Petti M, Corbisier T, Ribeiro A, Theophilo C, de Lima FP, Figueira R (2014) Bioaccumulation of potentially toxic trace elements in benthic organisms of Admiralty Bay (king George Island, Antarctica). Mar Pollut Bull 79:321–325

Malinga M, Szefer P, Gabrielsen G (2010) Age, sex and spatial dependent variations in heavy metals levels in the glaucous gulls (*Larus hyperboreus*) from the Bjørnøya and Jan Mayen, Arctic. Environ Monit Assess 169:407–416

Mansouri B, Babaei H, Hoshyari E (2012) Heavy metal contamination in feathers of western reef heron (*Egretta gularis*) and Siberian gull (*Larus heuglini*) from hara biosphere reserve of southern Iran. Environ Monit Assess 184:6139–6145

Martinez-Haro M, Taggart M, Mateo R (2010) Pb-al relationships in waterfowl feces discriminate between sources of Pb exposure. Environ Pollut 158:2485–2489

Mateo R, Lacorte S, Taggart M (2016) An overview of recent trends in wildlife ecotoxicology. Curr Trends Wildlife Res 1:125–150

Mathews T, Fisher N (2008) Trophic transfer of seven trace metals in a four-step marine food chain. Mar Ecol Prog Ser 367:23–33

Metcheva R, Yurukova L, Teodorova S, Nikolova E (2006) The penguin feathers as bioindicator of Antarctica environmental state. Sci Total Environ 362:259–265

Metcheva R, Yurukova L, Bezrukov V, Beltcheva M, Yankov Y, Dimitrov K (2010) Trace and toxic elements accumulation in food chain representatives at Livingston Island (Antarctica). Int J biol 2:155

Metcheva R, Yurukova L, Teodorova SE (2011) Biogenic and toxic elements in feathers, eggs, and excreta of gentoo penguin (*Pygoscelis papua ellsworthii*) in the Antarctic. Environ Monit Assess 182:571–585

Monteiro LR, Furness RW (2001) Kinetics, dose-response, and excretion of methylmercury in free-living adult Cory's shearwaters. Environ Sci Technol 35:739–746

Morera M, Sanpera C, Crespo S, Jover L, Ruiz X (1997) Inter- and intraclutch variability in heavy metals and selenium levels in Audouin's gull eggs from the Ebro Delta, Spain. Arch Environ Contam Toxicol 33:71–75

Nayak P (2002) Aluminum: impacts and disease. Environ Res 89:101–115

Neff JM (1997) Ecotoxicology of arsenic in the marine environment. Environ Toxicol Chem 16:917–927

Newman MC (2015) Fundamentals of ecotoxicology: the science of pollution. CRC Press, Boca Raton, FL, p 680

Nordberg M, Nordberg GF (2016) Trace element research-historical and future aspects. J Trace Elem Med Biol 38:46–52

Norheim G, Borch-Iohnsen B (1990) Chemical and morphological studies of liver from eider (*Somateria mollissima*) in Svalbard with special reference to the distribution of copper. J Comp Pathol 102:457–466

Nygard T, Lie E, Rov N, Steinnes E (2001) Metal dynamics in an Antarctic food chain. Mar Pollut Bull 42:598–602

Nyholm NE (1981) Evidence of involvement of aluminum in causation of defective formation of eggshells and of impaired breeding in wild passerine birds. Environ Res 26:363–371
O'Flaherty EJ (1998) Physiologically based models of metal kinetics. Crit Rev Toxicol 28:271–317
Ochoa-Acuña H, Sepúlveda MS, Gross TS (2002) Mercury in feathers from Chilean birds: influence of location: feeding strategy, and taxonomic affiliation. Mar Pollut Bull 44:340–345
Ohlendorf HM, Kilness AW, Simmons JL, Stroud RK, Hoffman DJ, Moore JF (1988) Selenium toxicosis in wild aquatic birds. Toxicol Environ Health 24:67–92
Orłowski G, Polechoński R, Dobicki W, Zawada Z (2007) Heavy metal concentrations in the tissues of the black-headed gull *Larus ridibundus* L. nesting in the dam reservoir in south-western Poland. Pol J Ecol 55:777–787
Outridge PM, Scheuhammer AM (1993) Bioaccumulation and toxicology of chromium: implications for wildlife. Rev Environ Contam Toxicol 130:31–77
Parslow JLF, Jefferies DJ, Hanson HM (1973) Gannet mortality incidents in 1972. Mar Pollut Bull 4:41–43
Pedro S, Xavier JC, Tavares S, Trathan PN, Ratcliffe N, Paiva VH, Medeiros R, Pereira E, Pardal M (2015) Feathers as a tool to assess mercury contamination in gentoo penguins: variations at the individual level. PLoS One 10:e0137622. doi:10.1371/journal.pone.0137622
Pérez-López M, Cid-Galán F, Hernández-Moreno D, Oropesa-Jiménez AL, López-Beceiro A, Fidalgo-Álvarez LE, Soler-Rodríguez F (2005) Contenido de metales pesados en hígado y plumas de aves marinas afectadas por el accidente del "Prestige" en la costa de Galicia. Rev Toxicol 22:191–199. (in Spanish)
Prashanth L, Kattapagari KK, Chitturi RT, Baddam VR, Prasad LK (2016) A review on role of essential trace elements in health and disease. J NTR Univ Health Sci 4:75–85
Raidal SR, Shearer PL, Cannell BL, RJDB N (2006) Micromelia in little penguins (*Eudyptula minor*). J Avian Med Surg 20:258–262
Rattner BA, Golden NH, Toschik PC, McGowan PC, Custer TW (2008) Concentrations of metals in blood and feathers of nestling ospreys (*Pandion haliaetus*) in Chesapeake and Delaware bays. Arch Environ Contam Toxicol 54:114–122
Ribeiro AR, Eira C, Torres J, Mendes P, Miquel J, Soares AMVM, Vingada J (2009) Toxic element concentrations in the razorbill *Alca torda* (Charadriiformes, Alcidae) in Portugal. Arch Environ Contam Toxicol 56:588–595
Rodrigue J, Champoux L, Leclair D, Duchesne JF (2007) Cadmium concentrations in tissues of willow ptarmigan (*Lagopus lagopus*) and rock ptarmigan (*Lagopus muta*) in Nunavik, northern Québec. Environ Pollut 147:642–647
Roth JA (2006) Homeostatic and toxic mechanisms regulating manganese uptake, retention, and elimination. Biol Res 39:45–57
Rothschild RFN, Duffy LK (2005) Mercury concentrations in muscle, brain and bone of western Alaskan waterfowl. Sci Total Environ 349:277–283
Sagerup K, Savinov V, Savinova T, Kuklin V, Muir D, Gabrielsen G (2009) Persistent organic pollutants, heavy metals and parasites in the glaucous gull (*Larus hyperboreus*) on Spitsbergen. Environ Pollut 157:2282–2290
Sanchez-Hernandez JC (2000) Trace element contamination in Antarctic ecosystems. Rev Environ Toxicol 166:83–127
Sánchez-Virosta P, Espína S, García-Fernández AJ, Eeva T (2015) A review on exposure and effects of arsenic in passerine birds. Sci Total Environ 512-513:506–525
Santos IR, Silva-Filho EV, Schaefer C, Maria S, Silva CA, Gomes V, Passos MJ, Van Ngan P (2006) Baseline mercury and zinc concentrations in terrestrial and coastal organisms of Admiralty Bay, Antarctica. Environ Pollut 140:304–311
Šaric M, Lucchini R (2007) Manganese. In: Nordberg GF, Fowler BA, Nordberg M, Friberg L (eds) Handbook on the toxicology of metals. Academic Press, London, pp 645–674

Savinov VM, Gabrielsen GW, Savinova TN (2003) Cadmium, zinc, copper, arsenic, selenium and mercury in seabirds from the Barents Sea: levels, inter-specific and geographical differences. Sci Total Environ 306:133–158

Scheifler R, Gauthier-Clerc M, Bohec CL, Crini N, Cœurdassier M, Badot PM, Giraudoux P, Maho YL (2005) Mercury concentrations in king penguin (*Aptenodytes patagonicus*) feathers at Crozet Islands (sub-Antarctic): temporal trend between 1966–1974 and 2000–2001. Environ Toxicol Chem 24:125–128

Scheuhammer AM (1987) The chronic toxicity aluminium, cadmium, mercury and lead in birds: a review. Environ Pollut 46:263–295

Scheuhammer AM, Basu N, Burgess NM, Elliot JE, Campbell GD, Wayland M, Champoux L, Rodrigue J (2008) Relationships among mercury, selenium, and neurochemical parameters in common loons (*Gavia immer*) and bald eagles (*Haliaeetus leucocephalus*). Ecotoxicology 17:93–101

Sjögren B, Iregren A, Elinder C-G, Yokel RA (2007) Aluminum. In: Nordberg GF, Fowler BA, Nordberg M, Friberg L (eds) Handbook on the toxicology of metals. Academic Press, London, pp 339–352

Skoric S, Visnjić-Jeftic Z, Jaric I, Djikanovic V, Mickovic B, Nikcevic M, Lenhardt M (2012) Accumulation of 20 elements in great cormorant (*Phalacrocorax carbo*) and its main prey, common carp (*Cyprinus carpio*) and Prussian carp (*Carassius gibelio*). Ecotoxicol Environ Saf 80:244–251

Smichowski P, Vodopivez C, Muñoz-Olivas R, Gutierrez AM (2006) Monitoring trace elements in selected organs of Antarctic penguin (*Pygoscelis adeliae*) by plasma-based techniques. Microchem J 82:1–7

Soria ML, Repetto G, Repetto M (1995) Revisión general de la toxicología de los metales. In: Repetto M (ed) Toxicología avanzada. Ediciones Díaz de Santos, Madrid, pp 293–358 (in Spanish)

Sparling DW, Lowe TP (1996) Environmental hazards of aluminum to plants, invertebrates, fish, and wildlife. Rev Environ Contam Toxicol 145:1–127

Steinhagen-Schneider G (1986) Cadmium and copper levels in seals, penguins and skuas from the Weddell Sea in 1982/1983. Polar Biol 5:139–143

Stewart FM, Phillips RA, Catry P, Furness RW (1997) Influence of species, age and diet on mercury concentrations in Shetland seabirds. Mar Ecol-Prog Ser 151:237–244

Suedel BC, Boraczek JA, Peddicord RK, Clifford PA, Dillon TM (1994) Trophic transfer and biomagnification potential of contaminants in aquatic ecosystems. Rev Environ Contam Toxicol 136:21–89

Sun L, Xie Z (2001) Changes in lead concentration in Antarctic penguin droppings during the past 3,000 years. Environ Geol 40:1205–1208

Szefer P, Pempkowiak J, Skwarzec B, Bojanowski R, Holm E (1993) Concentration of selected metals in penguins and other representative fauna of the Antarctica. Sci Total Environ 138:281–288

Szopińska M, Namieśnik J, Polkowska Z (2016) How important is research on pollution levels in Antarctica? Historical approach, difficulties and current trends. Rev Environ Contam Toxicol. doi: 10.1007/398_2015_5008

Tartu S, Goutte A, Bustamante P, Angelier F, Moe B, Clément-Chastel C, Bech C, Gabrielsen GW, Bustnes JO, Chastel O (2013) To breed or not to breed: endocrine response to mercury contamination by an Arctic seabird. Biol Lett 9:20130317. doi:10.1098/rsbl.2013.0317

Tartu S, Bustamante P, Angelier F, Lendvai AZ, Moe B, Blévin P, Bech C, Gabrielsen GW, Bustnes JO, Chastel O (2016) Mercury exposure, stress and prolactin secretion in an Arctic seabird: an experimental study. Funct Ecol 30:596–604

Thompson DR (1990) Metal levels in marine vertebrates. In: Furness RW, Rainbow PS (eds) Heavy metals in the marine environment. CRC, Boca Raton, FL, pp 143–182

Tin T, Fleming Z, Hughes K, Ainley D, Convey P, Moreno C, Pfeiffer S, Scott J, Snape I (2009) Impacts of local human activities on the Antarctic environment. Antarct Sci 21:3–33

UICN (2016) The IUCN Red List of Thereatened Species 2014.3. http://www.iucnredlist.org/details/22697817/0. Accessed 2 Jan 2016

Wastney ME, House WA, Barnes RM, Subramanian KN (2000) Kinetics of zinc metabolism: variation with diet, genetics and disease. J Nutr 130:1355–1359

Williams TD (1990) Annual variation in breeding biology of gentoo penguin, *Pygoscelis papua*, at Bird Island, South Georgia. J Zool 222:247–258

Yin X, Xia L, Sun L, Luo H, Wang Y (2008) Animal excrement: a potential biomonitor of heavy metal contamination in the marine environment. Sci Total Environ 399:179–185

Zamani-Ahmadmahmoodi R, Alahverdi M, Mirzaei R (2014) Mercury concentrations in common tern *Sterna hirundo* and slender-billed gull *Larus genei* from the Shadegan marshes of Iran, in north-western corner of the Persian Gulf. Biol Trace Elem Res 159:161–166

Zhang W, Ma J (2011) Waterbirds as bioindicators of wetland heavy metal pollution. Procedia Environ Sci 10:2769–2774

Tributyltin: Advancing the Science on Assessing Endocrine Disruption with an Unconventional Endocrine-Disrupting Compound

Laurent Lagadic, Ioanna Katsiadaki, Ron Biever, Patrick D. Guiney, Natalie Karouna-Renier, Tamar Schwarz, and James P. Meador

Contents

L. Lagadic (✉)
Bayer AG, Research and Development, Crop Science Division, Environmental Safety, Alfred-Nobel-Straße 50, Monheim am Rhein 40789, Germany
e-mail: laurent.lagadic@bayer.com

I. Katsiadaki • T. Schwarz
Centre for Environment, Fisheries and Aquaculture Science, Barrack Road, The Nothe, Weymouth, Dorset DT4 8UB, UK
e-mail: ioanna.katsiadaki@cefas.co.uk; namara1385@yahoo.com

R. Biever
Smithers Viscient, 790 Main Street, Wareham, MA 02571, USA
e-mail: rbiever@smithers.com

P.D. Guiney
University of Wisconsin-Madison, 777 Highland Avenue, Madison, WI 53705-2222, USA
e-mail: pdguiney@gmail.com

N. Karouna-Renier
USGS Patuxent Wildlife Research Center, BARC East Bldg 308, 10300 Baltimore Avenue, Beltsville, MD 20705, USA
e-mail: nkarouna@usgs.gov

J.P. Meador
Environmental and Fisheries Sciences Division, Northwest Fisheries Science Center, National Marine Fisheries Service, National Oceanic and Atmospheric Administration, Seattle, WA 98112, USA
e-mail: james.meador@noaa.gov

P. de Voogt (ed.), *Reviews of Environmental Contamination and Toxicology Volume 245*, Reviews of Environmental Contamination and Toxicology 245,
DOI 10.1007/398_2017_8

1 Introduction

Tributyltin (TBT) was introduced as a biocide in the 1960s and today its use is widely restricted by a variety of statutes. Although some countries (and local governments within those countries) restricted the use of TBT in antifouling paints as early as 1982 (Alzieu 2000; Champ 2000), it was not until 2008 that a more global ban was enacted. In that year, the International Convention on the Control of Harmful Antifouling Systems for Ships required its signatories to ensure that vessels would no longer use hull paint containing TBT or other organotin chemicals. This agreement was signed by 74 countries as of the end of October 2016. Unfortunately, a recent study has confirmed that recreational vessels sampled from countries around the Baltic Sea still contain high concentrations of TBT and triphenyltin (TPT) and may be a source to the environment (Lagerström et al. 2016).

The main inputs of tributyltin into the environment are contaminated water and sediment originating from ports, harbors, marinas, and boat yards due to leaching from boat paint and improper disposal. Contaminated sediment can be mobilized by dredging, bioturbation, ship scour, or weather events and TBT-contaminated water can be carried by currents to previously unimpacted locations. Tributyltin may also be introduced into the environment by ongoing release from previously treated structures (continuous release and during cleaning/renovation) and small-scale (mis-)use or disposal of prohibited antifouling products. Diffuse exposure may also arise from the use or disposal of previously TBT-treated wooden articles and other applications including its use as a stabilizer for PVC products, antifungal treatment, and as a preservative for wood, paper, and textiles (US EPA 2003).

Tributyltin may be one of the most toxic man-made chemicals ever intentionally released into the environment, eliciting endocrine-type responses at concentrations in the range of 1 ng/L for water or 10 ng/g for whole body. Even concentrations in the range of low μg/L aqueous or low μg/g tissue cause high rates of mortality in many species. However, outright mortality events due to TBT exposure are relatively rare. Therefore, the most important environmental consequences result from sublethal responses. Among them, the development of male sexual characteristics in female marine gastropods exposed to TBT, a phenomenon known as imposex, has been abundantly documented (e.g., Gibbs and Bryan 1996a). This abnormality has resulted in reproductive failure of populations of Caenogastropods globally, leading to mass extinction and subsequent alterations in community structure and functioning of coastal ecosystems (Gibbs and Bryan 1986, 1996b; Hawkins et al. 1994).

The main goal for this review was to examine the available literature on TBT as an EDC and provide a synopsis on population-relevant responses across major taxa. Additionally, we highlight the case of TBT as an unusual endocrine disruptor and discuss some of the reasons why its toxic potential and MeOA went unrealized for many years. Finally, we use the Organization for Economic Co-operation and Development (OECD) Conceptual Framework for Endocrine Disruptor Testing and Assessment to organize the available information on effect assessment and environmental exposure levels to conduct a tentative retrospective environmental risk assessment of TBT.

2 Methods

2.1 Literature Search and Selection of Data

To conduct this review, 160 references were selected from an initial list of approximately 965 regulatory reports and open and grey literature, in an attempt to capture relevant data from original studies on fish (45 references), molluscs (55 references), and other taxonomic groups including mammals (60 references); these were sorted according to the type of effect. This was not intended to be an exhaustive review of the extensive TBT literature, so it is possible that some relevant studies were inadvertently omitted.

2.2 Quality Evaluation of Relevant Data

TBT is a data-rich compound with numerous tests at various levels of biological organization. However, the vast majority of these studies did not follow international standardized test guidelines. Nevertheless, several full or partial life-cycle tests are available for mammals, fish, and several invertebrate taxa. Data previously validated for regulatory reviews (e.g., EU 2005; US EPA 2003, 2008) were assumed reliable; however, as far as possible, original studies were used as sources of data. Other ecotoxicity studies were quality checked and scored using Klimisch scores (e.g., using the ToxRTool: https://eurl-ecvam.jrc.ec.europa.eu/about-ecvam/archive-publications/toxrtool). Only studies ranked as Klimisch 1 and 2 were used for the subsequent analysis (although a small number of papers ranked as Klimisch 3 or 4 were used as supporting information if their findings were verified by other studies). However, Klimisch scores do not apply to field studies, which were a major source of data. Similarly, histopathological investigations are difficult to evaluate using these criteria. Therefore, expert judgment was used to evaluate the validity/credibility of these studies.

Many laboratory studies were performed using static or semi-static exposure regimes, and the reported effect concentrations were frequently based on nominal concentrations, due to a lack of analytical verification of the test concentrations over the study duration. TBT is adsorptive and test concentrations in non-flow-through studies are likely to be highly variable. This also means that equipment can easily be contaminated, thus affecting the actual exposure concentrations. In addition, a variety of units have been used to express the TBT concentration in the literature. To aid comparisons between studies, concentrations are expressed hereafter in a common unit (TBT ion). However, for simplicity and since the difference in molecular weight is small, any concentrations reported as TBTO (bis(tributyltin) oxide) or TBTCl (tributyltin chloride) were assumed to be effectively the same as TBT. Conversion to TBT was therefore made when the unit was originally expressed in terms of tin (Sn) and was accomplished by applying a factor of 2.44, which is the difference in molecular weight.

Other important limitations of the data included wide spacing between test concentrations and consequently between the LOEC and NOEC in some laboratory studies (very few report data in terms of EC_x values), poor methodological descriptions/statistical analyses, and uncertainties in the association between reported dissolved concentrations and observed effects in field studies (since the concentration at the time of adverse event initiation may be different, and organisms may accumulate TBT over a long period).

3 Environmental Fate and Occurrence

3.1 Physical and Chemical Properties

Several tributyltin compounds exist, but in general they all rapidly dissociate in water to form the tributyltin (TBT) cation, which is the toxic moiety. An example is bis(tributyltin) oxide ("tributyltin oxide" or TBTO), which has a vapor pressure of less than 0.016 Pa at 20 °C, a water solubility in the range 0.7–71 mg/L at 20 °C (depending on pH), and an octanol-water partition coefficient (K_{ow}) around 3.5 (as a log value) (e.g., ECHA 2008; EU 2005; US EPA 2003). The pH-specific K_{ow} (D_{ow}) for TBT is constant above pH 6 (Arnold et al. 1997). An organic carbon-water partition coefficient of 32,000 ($\log_{10} K_{oc} = 4.5$) was proposed by Meador (2000).

3.2 Uses

Historically, TBT compounds were widely used as biocides in antifouling products, parasite control products, and wood preservatives. Regulatory controls have been implemented in many jurisdictions (e.g., under the International Maritime Organization's Anti-Fouling Systems Convention), but some residual biocidal uses in these and other types of industrial and consumer products (including sports clothing) may still occur in some parts of the world (Antizar-Ladislao 2008; Choudhury 2014). They are also used as chemical intermediates for the production of other organotins (e.g., dibutyltin stabilizers for PVC), and may therefore occur as unintentional impurities (e.g., OECD 2007). However, these uses are declining.

3.3 Metabolism

TBT can be metabolized sequentially by cytochrome P450 (CYP450) enzymes to dibutyl- and monobutyltin (e.g., Cooke et al. 2008; Strand et al. 2009). Metabolic capabilities vary widely among taxa (e.g., Ohhira et al. 2003, 2006a, b; Bartlett et al. 2007; Oehlmann et al. 2007; Yang et al. 2009) and even between sexes in mammals (Ohhira et al. 2006b). High bioaccumulation levels in invertebrates and fish are believed to be due to a low capacity for metabolism/excretion and high rates of uptake (Meador 1997).

3.4 Potential Exposure Routes

TBT has a degradation half-life of days to months in water and up to several years in sediment (e.g., ECHA 2008). It is very bioaccumulative, with whole fish bioconcentration factors in the range 2000–50,000 L/kg. The bioaccumulation potential in molluscs can be higher, but is generally similar to that for fish (e.g., Meador 2006; ECHA 2008). TBT bioaccumulation does not follow equilibrium partitioning (Meador 2000). Consequently, some fish exhibit high bioaccumulation factors, which may result from high rates of ventilation. Aquatic organisms can be exposed via both the water column and ingestion of contaminated food (including sediment), although there are no data to suggest that biomagnification occurs in food webs. Terrestrial organisms may also be exposed via TBT-contaminated sediments (e.g., during flood events or disposal of sediment dredging), application of biocidal products and/or contaminated sewage sludge to soil, atmospheric deposition, and by ingestion of contaminated food or water (Antizar-Ladislao 2008; Silva et al. 2014).

3.5 Environmental Concentrations

For this review we did not conduct an exhaustive search for environmental concentrations, but examined past compilations. Several review articles describe the occurrence of butyltins in water from locations around the world over the past several decades (Fent 1996; Antizar-Ladislao 2008). More recent monitoring performed in English and Welsh estuarine surface waters during 2012–2014 was selected as representative of current aquatic levels in a region known to be previously contaminated (UK Environment Agency, pers. com. 2016). The dataset has a number of non-detects; however, the following information is considered to be reasonably conservative for aqueous concentrations: 90th percentile: 0.5 ng/L; arithmetic mean: 0.3 ng/L; median: 0.2 ng/L; range: 0.1–8 ng/L ($n = 269$; two outlier values, 44 and 1368 μg/L, were removed prior to the calculation).

A recent review of tissue concentrations for biota from a variety of countries indicated relatively high TBT concentrations in molluscs, fish, aquatic birds, and marine mammals ranging from low ng/g to low μg/g concentrations, although there were indications of significant decreases with time in some species (Elliott et al. 2007; Mizukawa et al. 2009; Meador 2011).

4 Primary Molecular Initiating Event (MIE)

4.1 In Vitro and In Silico Analyses

Binding to a member of the nuclear receptor superfamily is a common MIE associated with endocrine and metabolic pathways. All relevant in vitro assays currently utilize mammalian receptors, assuming interspecies conservation of the structural and functional aspects of each receptor. Among those methods, ToxCast™ uses high-throughput screening methods and computational toxicology approaches to rank and prioritize chemicals. For the US EPA's Endocrine Disruption Screening Program (EDSP), the initial focus of these screening methods has been on estrogen, androgen, and thyroid (EAT) hormone interactions. Analysis of TBT using ToxCast™ identified activity in ER and AR assays at levels generally in excess of the cytotoxicity limit (Fig. 1). In addition, the EDSP21 Dashboard identified TBT as inactive for EAT screens (http://actor.epa.gov/edsp21/).

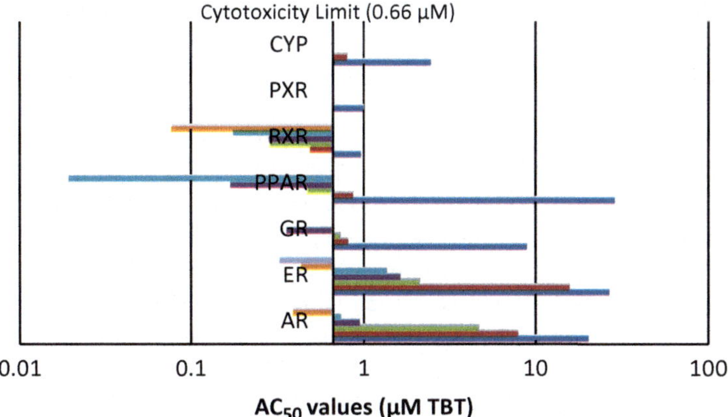

Fig. 1 Selected ToxCast™ nuclear receptor family AC_{50} data for TBT (downloaded March 2016). *CYP* Cytochrome P450, *PXR* pregnane X receptor, *RXR* retinoic X receptor, *PPAR* peroxisome proliferator-activation receptor, *GR* glucocorticoid receptor, *ER* estrogen receptor, *AR* androgen receptor

However, TBT activity in RXR and PPAR assays typically occurred at levels less than the cytotoxicity limit and at lower levels than most ER or AR assays. Thus, ToxCast™ identified the potential of TBT to act through activation of RXR and/or PPAR pathways, although TBT may have been missed as a potential EDC if screening was limited to EAT-related assays alone.

Using an RXR-permissive PPARγ reporter cell line, Grimaldi et al. (2015) demonstrated that, among other butyltins, TBT is able to activate RXR at nanomolar concentrations. Structural profiling using in silico 3D-modelling of the ligand-binding pocket (LBP) of the mollusc *Lymnaea stagnalis* RXR showed that amino-acid residues involved in the binding of RXR ligands (organotins, including TBT, and cis-9-retinoic acid) are identical between *Lymnaea* and humans (Boulahtouf et al. 2015). In addition, the RXR receptor from the freshwater mollusc *Biomphalaria glabrata* does not only bind retinoic acid, but also activates transcription (Bouton et al. 2005). Other studies on invertebrate RXRs have identified significant differences in the LBP as well as a number of mutations that result in low or no affinity of RXR for their vertebrate ligands (all retinoic acids). These include the retinoic acid receptor (RAR) from the marine snail *Nucella lapillus* (Gutierrez-Mazariegos et al. 2014) and the RXR from *Daphnia magna* (Wang et al. 2007), highlighting a knowledge gap regarding the natural ligands relevant to RXR signaling pathways in invertebrate species. We anticipate that with the increasing genomic resources for these important phyla, both the structure and the function of nuclear receptors such as RXR can be further elucidated. This will aid not only our understanding of evolutionally processes but also the risk posed by chemicals such as TBT, which can activate these important transcription factors.

4.2 Weight-of-Evidence for RXR and PPAR$_\gamma$ Pathways

4.2.1 Molluscs

From both in vivo (including injection experiments) and in vitro (transactivation assays) studies, there is strong evidence that TBT interacts with RXR in both marine (e.g., Castro et al. 2007) and freshwater (Boulahtouf et al. 2015) gastropod species. Such interactions were observed at TBT concentrations of 1 nM TBT (equiv. 290 ng TBT/L) in transactivation assays and 1000 ng TBT/g body wet weight through injection. It has also been shown that RXR gene transcription is modulated by TBT, with an increase in the penis-forming area of imposexed females (Lima et al. 2011). Hence, interaction of TBT with RXR seems to be the main molecular event initiating changes in the development of sexual organs in female snails, ultimately resulting in imposex. Importantly, imposex can be induced by cis-9-RA, the natural ligand of RXR (Nishikawa et al. 2004; Castro et al. 2007). More recently, the use of an open transcriptomic approach supported the involvement of steroid, neuroendocrine peptide hormone dysfunction and retinoid mechanisms in imposex induction by chronic exposure to TBT in *Nucella lapillus* (Pascoal et al. 2013). This study also suggested the involvement of PPAR pathways and showed that rosiglitazone, a well-known vertebrate PPAR$_\gamma$ ligand, induced imposex. Nevertheless, although it is certain that activation of RXR and/or RXR/PPAR is the MIE, the mechanistic links with subsequent pathways remain largely unexplained.

However, other primary molecular mechanisms for TBT effects have also been proposed. These include the activation of gonadotropin releasing hormone receptor (GnRHR) along with the gonadotropin releasing hormone (GnRH) (Castro et al. 2007). In *Octopus vulgaris*, it has been shown that a GnRH-like peptide contributed in vitro to an elevation of basal steroidogenesis of testosterone, progesterone, and 17β-oestradiol, in a concentration-dependent manner in both follicle and spermatozoa (Kanda et al. 2006). Another hypothesis formed around the imposex phenomenon suggested that TBT is a neurotoxicant, causing the aberrant secretion of neurohormones, primarily the neuropeptide APGWamide, which regulates male sexual differentiation in molluscs (Oberdorster and McClellan-Green 2002). However, the body of literature is not large enough to allow full evaluation of these additional or alternative pathways.

For many years, the mechanism of TBT-induced imposex was dominated by the steroid hypothesis. This was primarily due to researchers measuring high levels of free testosterone in tissues of impacted molluscs. The link between free testosterone and penis formation was made and supported by numerous publications (e.g., Oehlmann et al. 2007). Despite extensive experimental efforts, treatment with either testosterone or fadrozole (a potent aromatase inhibitor) did not replicate this condition (e.g., Iguchi and Katsu 2008). Although it is well established that TBT does affect many CYPs (including those involved in vertebrate steroidogenesis) and other metabolizing pathways (e.g., esterification), these effects alone do not constitute evidence of a physiological role for these steroids in molluscs. In fact, both the origin and physiological role of sex steroids in molluscs are still controversial (for reviews see Scott 2012, 2013). Current data suggest that they are likely to be accumulated from the environment. Gooding and Leblanc (2001) provided the first evidence on the ability of molluscan species to take up steroids from water, which was further supported by additional studies (e.g., Janer et al. 2004). More recently, a comprehensive evaluation for steroid uptake by both gastropod snails and bivalve molluscs (Giusti 2013; Giusti et al. 2013a; Schwarz 2015; Schwarz et al. 2017a, b) revealed that this process is very rapid, does not seem to have a saturation limit and is particularly strong for testosterone. Following steroid uptake from the environment, molluscs appear to esterify them (Giusti and Joaquim-Justo 2013) and store them as fatty acid esters through the action of acyl-CoA: testosterone acyl transferase (ATAT) (Janer et al. 2004: Gooding and Leblanc 2001). The retention of steroids as fatty acid esters appears to persist as there is little to no depuration, particularly of estradiol, testosterone, and progesterone from either snails or bivalves placed in clean water for up to 10 days (Schwarz 2015; Schwarz et al. 2017a, b).

Inhibition of either CYPs or phase II metabolism by TBT inevitably will reduce steroid clearance and metabolism, leading to an increase in free steroids. It has been suggested that TBT acts by reducing the retention of testosterone as fatty acid-esters, thus increasing the levels of free testosterone. This may play a role in the development of male sexual organs in females (LeBlanc et al. 2005). However, LeBlanc et al. (2005) added that two assumptions must be met before this putative causative relationship between TBT, testosterone, and imposex can be accepted. First, it must be accepted that testosterone is a male sex-differentiating hormone in molluscs and second, TBT specifically targets some component of the testosterone regulatory machinery causing the aberrant accumulation of this hormone in the snails. None of these has been proven to date. The lack of a nuclear AR or AR-like homologues in the genomes of molluscan species studied to date supports the lack of a physiological role for androgens in these species (Kaur et al. 2015; Vogeler et al. 2014).

4.2.2 Fish

The MIE for reproductive and metabolic impairment in fish is also expected to occur via the RXR and/or PPAR receptors. Lima et al. (2015) exposed zebrafish to only one dose each of TBT, cis-9-RA, and all-trans-retinoic acid in the diet. TBT at this one dose (2.4 μg/g in diet) affected fish weight, fecundity, and sex ratio; however, the natural RXR ligand (cis-9-RA) at 5 μg/g in diet did not. Zhang et al. (2013b) reported significant activation of RXRα in male rockfish (*Sebastiscus marmoratus*) brain when exposed to TBT at 2.4, 24, and 244 ng TBT/L. Female rockfish exhibited the opposite pattern with significant reductions in expression at the two highest doses. No differences in expression of $PPAR_\gamma$ were found for males or females at any dose. Several studies reported TBT-induced effects on adipogenesis via RXR/PPAR activation (Meador et al. 2011; Tingaud-Sequeira et al. 2011; Ouadah-Boussouf and Babin 2016).

How exactly the RXR/PPAR activation affects reproduction in fish however is poorly understood. Diverse evidence point to the existence of RXR/PPAR and ER cross-talk. It has been shown however that PPAR and RXR, or their heterodimer, can bind directly to estrogen responsive elements (EREs) in the gene promoters (Nunez et al. 1997). Other studies suggest that RXR/PPAR activation directly affects CYP expression, including aromatase (Cheshenko et al. 2008). A PPAR/RXR responsive element was predicted in the zebrafish CYP19β promoter (Kazeto et al. 2001) whilst a RAR binding region was identified in the tilapia CYP19 promoter (Chang et al. 2005). This was further supported by in vitro studies with mammalian cell lines where TBT binding to RXR and PPAR leads to modulations of CYP19 expression (Nakanishi et al. 2005, 2006).

4.2.3 Mammals

As described above, Kanayama et al. (2005) presented some of the first evidence of the implication of the RXRα and $PPAR_{\Upsilon}$ pathways in TBT-mediated endocrine effects in mammals, and also proposed that this pathway is the likely route for the low-dose imposex response in gastropods. Additional studies (Nakayama et al. 2005; Grün et al. 2006; Le Maire et al. 2009) provided further data showing that TBT exerts its biological effects via transcriptional regulation of gene expression through activation of these receptors, implicating this pathway in the adipogenic effects of TBT. Mammalian aromatase gene expression is also regulated through the RXRα/$PPAR_{\Upsilon}$ pathway by various ligands, including TBT. However, the direction of the regulation in response to TBT appears to be dose- and tissue-specific, potentially because aromatase is regulated through tissue-specific promoters (Simpson et al. 1993).

4.2.4 Summary

Retinoic acid binds to both RAR and RXR regulating the transcription of 500 genes involved in a large array of biological processes (Blomhoff and Blomhoff 2006). RA has long been recognized as a morphogen, important for axial patterning and organ formation in developing vertebrates. An adverse outcome pathway for neural tube and axial defects modulated by retinoic acid in vertebrates (including man) has been proposed (Tonk et al. 2015). This analysis was based on data from rat whole embryo culture, embryonic stem cells and the zebrafish embryotoxicity test, and identified certain conserved pathways on RA signaling between mammals and fish. Adverse effects in vertebrates included craniofacial and limb malformations/defects, which suggests that the TBT-induced developmental abnormalities in mollusks, such as shell abnormalities (Alzieu et al. 1986), and imposex are different expressions of disruption of the same pathway. Analysis of genomic data revealed that the important morphogenic role of RA does not only extend to invertebrate chordates (tunicates and cephalochordates), but also to other invertebrate groups, such as hemichordates and sea urchins (Campo-Paysaa et al. 2008).

Altogether evidence suggests that most of the endocrine effects of TBT have their origin in RXR and RXR/PPAR activation. An important determinant of the severity and magnitude of responses in different species is the structure of the ligand binding domain, as studies have demonstrated the evolutionary plasticity of this domain, whilst the function of RAR, RXR, and PPAR appears to be largely conserved.

5 Toxic Effects Plausibly Mediated by Endocrine Disruption

5.1 The OECD Conceptual Framework for Endocrine Disruptor Testing and Assessment

As previously mentioned, few studies were performed according to standard test guidelines that correspond to the OECD Conceptual Framework for Endocrine Disruptor Testing and Assessment (CFEDTA; see Table 1 of the Annexes) (OECD 2012a, b). Non-standard studies do not necessarily fit within the different levels of this framework. Hereafter, the studies with TBT are therefore organized according to the test design, including the exposure duration and portion of the life-cycle exposed, and to the type of biological responses. As EC_x were not always provided or could not be deduced from the study, endpoints are mainly reported as LOEC and/or NOEC.

Standard ecotoxicology data characterizing acute toxicity endpoints were not reviewed in any great detail for this review because overt toxicity did not impact the evaluation of endpoints relevant to endocrine disruption. Acute toxicity for sensitive species was generally one to two orders of magnitude greater than LOECs for the relevant chronic apical and endocrine disruption endpoints.

5.2 Non-test Approach: Summary of Toxicological Information

The level 1 of the OECD CFEDTA corresponds to non-test information that can be used to define the general toxicological profile of a chemical with respect to its endocrine disruption properties (OECD 2012a). QSAR analyses have been conducted in some organisms (e.g., green algae; Neuwoehner et al. 2008). However, they only provide patchy information and do not appropriately cover taxa of interest such as molluscs or fish. Therefore, in the context of the present review, TBT has been evaluated in the US EPA's ToxCast™ program (http://epa.gov/ncct/toxcast/). It was classified as being a "promiscuous" chemical because it exhibited

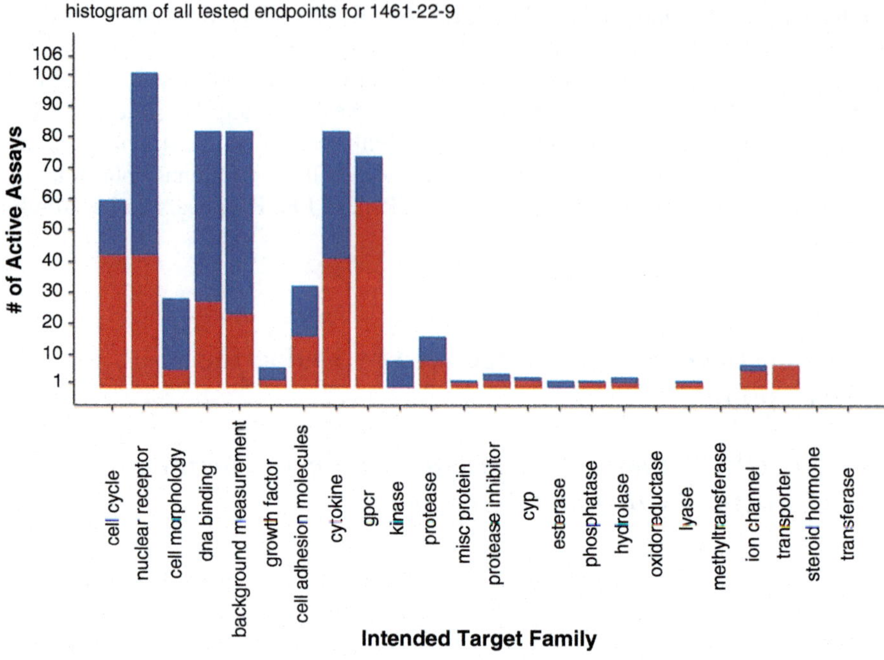

Fig. 2 Result of the ToxCast™ analysis for TBT (*red* = active assays and *blue* = inactive assays) (Downloaded March 2016)

activity in approximately 285 assays across 20 target families (Fig. 2). This type of profile suggests that TBT will act across a number of toxic pathways potentially including endocrine and non-endocrine molecular initiating events.

5.3 Invertebrates

5.3.1 In Vitro Assays

In molluscs, few studies have reported the use of in vitro assays for identifying initial endocrine mechanisms (i.e., MIE) due to TBT exposure. These studies employed transactivation assays using mammalian cells transfected with RXR from *Thais clavigera* (Urishitani et al. 2013) and *Lymnaea stagnalis* (Boulahtouf et al. 2015). The results demonstrated that TBT binds the LBD of RXR in these species, and its affinity is equivalent to that of 9-cis-RA, the natural ligand of RXR.

5.3.2 Physiological Responses

In the reproductive tissues of female dog whelk *N. lapillus*, RXR gene transcription is increased in the penis-forming area associated with the formation of a penis and/or vas deferens (Lima et al. 2011). In this species, 9-cis-RA is as potent as TBT in inducing imposex, indicating that TBT toxicity in gastropods is mediated through the modulation of the RXR signaling pathways (Nishikawa et al. 2004; Castro et al. 2007). It should be noted that imposex is also induced by rosiglitazone, a PPARγ agonist, suggesting that the heterodimer of RXR-PPAR is a critical pathway for this phenomenon, at least in *N. lapillus* (Pascoal et al. 2013). This study also suggested that other transcription factors such as LXR and lipophilic orphan receptors are involved in the TBT toxicity pathway.

Interestingly, TBT appears to affect the endocrine system of insects as well (Hahn and Schulz 2002). In an in vivo study on *Chironomus riparius*, an environmentally relevant concentration of TBT (1 ng/L) showed effects on many endocrine-related genes, including up-regulation of the ecdysone receptor, the ultraspiracle gene (the orthologue of RXR in insects), the estrogen-related receptor, and the E74 early ecdysone inducible gene, whilst the vitellogenin (Vtg) gene remained unaffected (Morales et al. 2013). The same study showed genotoxic effects of TBT in insects by means of the comet assay.

5.3.3 Organismal Effects

A wide variety of organ responses are known in molluscs exposed to low concentrations of TBT. These include penis development in female snails, abnormal testis histopathology, and sperm alteration (count, motility, abnormality). These responses occur in the 1–10 ng TBT/L range for aqueous concentrations in the environment (e.g., Horiguchi et al. 1994; Leung et al. 2006) and 10 to 100 ng TBT/g wet weight (ww) whole-body tissue (Meador 2011) in affected gastropods. Female prosobranch molluscs exposed in the laboratory can develop imposex with LOECs of 1 to 83 ng/L (Gooding et al. 2003; Abidli et al. 2012, 2013).

In gastropods, adverse effects resulting from in vivo exposure to TBT mainly concern reproductive impairment associated with alterations of the sexual organs in females, including the staged development of penis and vas deferens known as imposex. The Vas Deferens Sequence Index (VDSI) has been developed to characterize the extent of imposex in *N. lapillus* and *T. clavigera* (e.g., Gibbs et al. 1987; Blackmore 2000; Leung et al. 2006). VDSI has seven stages, with stage 0 indicating no imposex and stage 6 indicating female sterilization. Several studies indicate that the imposex threshold concentration lies at 1 ng TBT/L, with increasing sterilization as concentrations increase (e.g., Gibbs et al. 1988). Gibbs (1996) found that juvenile female *Ocenebra erinacea* exposed to 7.3 ng TBT/L developed a longitudinal split of the oviduct wall. Adult *O. erinacea* collected from TBT-contaminated sites with advanced imposex exhibited the same lesions and through laboratory spawning experiments were found to be sterile (no capsules produced). Significant reductions in gastropod reproduction were generally found at levels slightly higher than those that induced imposex, with effective concentrations ranging from 12 to 1000 ng TBT/L (Duft et al. 2007; Leung et al. 2007; Giusti et al. 2013a, b). Other invertebrate species are known to exhibit reproductive effects across a wide aqueous concentration range of 10 to 2225 ng TBT/L (Oberdorster et al. 1998; Ohji et al. 2003a, b; Huang et al. 2010).

Developmental toxicity of TBT was also investigated, using a 21-day embryo test with the freshwater snail *Lymnaea stagnalis*. TBT (added as TBTCl) had NOEC values of 30 ng TBT/L and 100 ng TBT/L for the mean hatching time and hatching success, respectively (Bandow and Weltje 2012). These values compare well with the NOEC of 105 ng TBT/L obtained for the fecundity of adult *L. stagnalis* exposed to TBT hydride for 21 days (Giusti et al. 2013a, b).

5.3.4 Life-Cycle Studies and Population-Level Responses

Limited full life-cycle or multigenerational laboratory studies using TBT have been conducted with invertebrates other than molluscs. The calanoid copepods *Pseudodiaptomus marinus* and *Schmackeria poplesia* exhibited similar sensitivities, with LOECs of 6 and 20 ng TBT/L, respectively (Huang et al. 2006, 2010). The amphipod *Caprella danilevskii* exhibited similar sensitivity to the copepods with a LOEC of 10 ng TBT/L (Ohji et al. 2003a, b). *Daphnia magna* was thought to be much less sensitive to TBT, with a LOEC of 2225 ng TBT/L in a two-generation study (Oberdorster et al. 1998). However, a recent study reported reduced brood size, total offspring, neonate volume, and neonate length at 88 ng TBT/L, with many of these effects being observed in the F1 and F2 generations (Jordão et al. 2015). Altered lipid homeostasis was suggested to be responsible for the abnormalities in reproduction.

For gastropods exposed to TBT, life-cycle and population studies are mainly represented by long-term laboratory exposures (>1 year) or field monitoring. In gastropods, a large number of studies reported decline to extinction for populations with increasing proportions of imposexed females (Gibbs 1996, 2009). Several studies indicate that TBT in the marine environment can impact populations of Caenogastropod snails through female sterilization associated with imposex (Spence et al. 1990; Bailey et al. 1995; Harding et al. 1997). These population-level responses were associated with water concentrations in the range 1–10 ng TBT/L, which is consistent with molecular studies characterizing the affinity of TBT for the RXR-PPAR receptor, in addition to other known ligands. It is important to note that there is not always a linear relationship between imposex development, as assessed through the VDSI, and female sterility (Barroso et al. 2002). On the other hand, female gametogenic activity occurs with no apparent differences in imposex-affected and imposex-free populations (Avaca et al. 2015). This may have important implications for population recovery.

Shell development in molluscs may be as sensitive as imposex and reproduction with a number of species exhibiting shell growth and development effects in aqueous concentrations ranging from 8 to 1000 ng TBT/L (Leung et al. 2006, 2007; Giusti et al. 2013b). Shell abnormalities in oysters were reported as early as the beginning of the 1980s (Alzieu et al. 1986), showing that oyster populations have also been impacted by TBT. For example, a correlation was found between rock oyster (*Saccostrea glomerata*) population density and the discontinued use of TBT in estuaries with high densities of boat moorage (Birch et al. 2014). This conclusion is also supported by other studies that reported declining tissue concentrations in molluscs from this area over the same time period (Batley et al. 1992; Lewis et al. 2010).

5.3.5 Reversibility of the Effects

The degree of imposex reversibility in molluscs depends on the species. Once TBT exposure stops, female penis length declines slowly in *Nassarius reticulatus* (Bryan et al. 1993), and more quickly in female *Ilyanassa obsoleta* (Smith 1981). Contrary to this, imposex in *N. lapillus* is largely irreversible (Bryan et al. 1987). Population recovery can therefore be slow, especially for species that are long-lived and/or for which recruitment is limited, such as *N. lapillus* (Matthiessen and Gibbs 1998; Oehlmann et al. 2007; OECD 2010) and *N. reticulatus* (Couceiro et al. 2009). Nevertheless, there appears to be a widespread amelioration worldwide (e.g., Canada, US EPA 2003; Hong-Kong, Leung et al. 2006; Spain, Couceiro et al. 2009; England and Wales, Nicolaus and Barry 2015) as populations of snails have recovered significantly after the reduction in use of TBT as an antifoulant (Birchenough et al. 2002; Bray et al. 2012; Nicolaus and Barry 2015).

5.4 *Fish*

5.4.1 In Vitro Assays

Organotins, including TBT, are well-known inhibitors of the hepatic microsomal CYP450 systems in a variety of fish species (review by Fent and Hunn 1996). Protein transcription, enzyme activity, and reductases are all affected. The isoform CYP450 1A1 appears to be particularly sensitive to TBT exposure. Other CYP450 isoforms, including those with testosterone hydroxylase activity, are also inhibited, albeit at high concentrations. In general, CYP450 inhibition occurs at levels that are close to cytotoxicity and as such it is difficult to establish whether the inhibitions observed constitute an additional mechanism of toxicity, or they are indeed an effect of general toxicity affecting a variety of sulfhydryl-containing proteins. It has since been established that not only phase I but also phase II metabolism enzymes are affected by TBT (Morcillo et al. 2004). It is interesting to note that in this study, glucuronidation of testosterone but not estradiol was inhibited by incubating fish liver microsomes with TBT at concentrations as low as 5 μM.

The general inhibitory effect of moderate to high doses of TBT on hormonal and biotransformation pathways has also been confirmed in salmon hepatocytes using gene expression patterns (Vtg, ER, AR) and CYP-mediated enzyme activities as endpoints (Mortensen and Arukwe 2009). These consistent decreases in cellular responses over time and with increasing TBT concentrations suggest a possible inhibitory effect of TBT on transcription. The effect of TBT on other transcription factors such as RXR and PPAR in fish cell systems is surprisingly under-studied. Nevertheless, TBT inhibited plaice (*Pleuronectes platessa*) $PPAR_{\alpha}$ and $PPAR_{\beta}$ at 1 nM in transfection assays although it had no effect on $PPAR_{\gamma}$ (Colliar et al. 2011).

5.4.2 Physiological Responses

Several studies have examined various physiological responses in fish exposed to TBT. A series of studies have been published on TBT-induced effects on rockfish (*Sebastiscus marmoratus*) gonadal development. In the first study, Zhang et al. (2007) brought wild female fish into captivity for experimentation. Ovarian testosterone levels of fish exposed to 10 ng TBT/L significantly increased, whereas levels in ovaries exposed to 100 ng/L did not significantly change compared to the control. Exposure to 10, 100 ng/L TBT resulted in significantly decreased 17β-estradiol levels in the ovaries. No masculinization was observed but the exposure period, limited to 50 days, might have been too short or the sexual differentiation stage as female was too advanced. Histological examination reportedly showed developmental suppression and atresia of ovarian follicles. However, these results may be unreliable as they included incorrect criteria to judge the follicles development stage [e.g., follicles labeled vitellogenic were not vitellogenic, and the results of Terminal deoxynucleotidyl transferase (TdT) dUTP Nick-End Labeling (TUNEL) assay staining for apoptosis were not convincing; J. Wolf, pers. com. 2016]. Likewise, there was no mention of methods used to minimize observational histopathological bias.

In a second experiment (Zhang et al. 2009a, b), fish were exposed to nominal concentrations of 2.44, 24.4, and 244 ng TBT/L for 48 days. There was strong evidence of inhibition of testicular development, interstitial fibrosis and increased testicular lipid but no NOEC (only a LOEC) was established. Histopathological testicular changes were observed that may represent a stress response as opposed to specific endocrine or toxicological activity (Zhang et al. 2009a). Inhibition of thyroidal status related to depression of testicular development was also studied at TBT concentrations of 2.44, 24.4, and 244 ng TBT/L for 50 days. The NOEC was 2.44 ng TBT/L based on histopathology (testis: decreased spermatozoa, pyknosis, interstitial fibrosis; thyroid: decreased colloid). There was a possible effect of decreased spermatozoa at 244 ng TBT/L, and damage to the thyroid gland and a decrease in production of thyroid hormones were observed (T4 in serum significantly correlated with GSI) (Zhang et al. 2009b).

In a third study (Zhang et al. 2011), TBT-induced RXR_{α} expression in embryos at 0.1 and 1 ng/L, an effect thought to be responsible for the induction of increasing apoptotic cells. TBT was also shown to induce ectopic lipid accumulation in ovarian interstitial cells (consistent with PPAR activation) and decreased testosterone esterification in the ovaries leading to increased free testosterone (NOEC = 1 ng/L).

In a final study (Zhang et al. 2013a, b), 30 male and 30 female fish per group were exposed to nominal concentrations of TBT of 1, 10, and 100 ng/L. CYP19b expression in the male fish significantly increased ($p = 0.026$, 0.04, and 0.02, respectively) after exposure to TBT, the highest elevation being 3.12-fold in the 10 ng/L group. In the female fish, the CYP19b expression increased slightly in the 10 and 100 ng/L groups, but this was not statistically significant ($p = 0.078$ and 0.234, respectively). Testosterone and estradiol levels were unchanged in males but testosterone increased and estradiol decreased in females. ER expression was affected in males (up-regulated at low concentrations; no difference at 100 ng/L) but not in females. RXR_{α} expression increased in males but decreased in females.

TBT is a complex endocrine disrupter in zebrafish (*Danio rerio*). McGinnis and Crivello (2011) injected TBT at 1–5 mg/kg intraperitoneally into fish, which directly inhibited ER-regulated processes by acting as a non-competitive inhibitor. TBT did not inhibit AR-regulated processes, but decreased acyl-transferases and sulfation of testosterone in the liver. TBT had an androgenizing effect in the brain but a feminizing effect in the liver and gonads. Rapid metabolization of TBT to di- and monobutyltin also occurred in the liver, resulting in complex and non-elucidated interactions with steroid pathways (McGinnis et al. 2012).

In the brown trout (*Salmo trutta fario*), TBT exposure showed a decreasing trend of ovarian CYP19 expression, but not a potent effect at 1000 ng/L (Pereira et al. 2011a, b), suggesting that TBT interferes with the steroidogenic pathway at a transcriptional level. The in vivo down-regulation of IGF2 in the pre-vitellogenic ovaries might indicate that TBT interferes with factors that are absent in the ex vivo gonad cultures (Pereira et al. 2011a, b). TBT did not affect testosterone or estradiol concentrations, further supporting previous evidence that the CYP19 modulating effects of this chemical are not mediated through direct inhibition of CYP19 activity (Pereira et al. 2011a, b). Aromatase expression in the brain, reproductive behavior, and secondary sexual characteristics were studied by Tian et al. (2015) in guppies (*Poecilia reticulata*). TBTCl treatment inhibited gene expression of CYP19A and CYP19B in brain of males, which led to altered reproductive behavior with a LOEC of 4.45 ng TBT/L.

A series of studies with juvenile Atlantic salmon (*Salmo salar*) that were force-fed TBT alone and in combination with forskolin, reported a number of affected gene expression patterns, including CYP3a, CYP11b, CYP19a, SF-1, glucocorticoid receptor, ERα PXR, PPARs, glutathione *S*-transferase (GST), ACOX 1, IL-b, TNFa, IFNγ, IFNα, Mx3, IGF-1, IL-10, and TGFb (Kortner et al. 2010; Pavlikova et al. 2010, Pavlikova and Arukwe 2011). Forskolin activates the enzyme adenylyl cyclase and increases intracellular levels of cAMP, an important second messenger necessary for the proper biological response of cells to hormones and other extracellular signals. Since most effects observed after TBT exposure were modulated by forskolin exposure, these studies suggested that TBT may exert its endocrine, biotransformation and lipid peroxidation effects via the cAMP/PKA second messenger system.

5.4.3 Organismal Effects

In contrast to the extensive literature dealing with the adverse impacts of TBT in mammalian and molluscan species, relatively few studies have addressed higher-level effects of TBT on fish. Bentivegna and Piatkowski (1998) studied embryotoxicity of TBT in medaka *Oryzias latipes* exposed to nominal concentrations of TBT acetate (TBTA). Results showed that 415 and 4150 ng/L produced 100% lethality in all age groups, while 41.5 ng/L produced no acute lethality in 3- and 5-day embryos, and between 16 and 33% lethality in 0-day embryos. Subchronic endpoints showed that toxicity was concentration-related and that embryos exposed on day 0 were more sensitive than those exposed on days 3 and 5. LOECs for hatching success were 10,440 ng/L for day 0 and 41,500 ng/L for days 3 and 5. LOECs for the combined effects of hatching success and gross abnormalities were 10,440 ng/L for day 0 and 20,590 ng/L for days 3 and 5. Although no endocrine-sensitive endpoints were measured, there was some evidence of reduced CYP450 induction.

It has been reported that TBT can alter the sex ratio towards males in zebrafish (McAllister and Kime 2003; Santos et al. 2006) and Japanese flounder (*Paralichthys olivaceus*) (Shimasaki et al. 2003). Genetic female flounder exhibited an increased rate of sex reversal when fed TBT in their diet. The proportion of males significantly increased to 25.7% in the 0.1 μg/g group and to 31.1% in the 1.0 μg/g group compared with the control (2.2%). Histological observations showed that, in the TBT-treated groups, normal females had typical ovaries and assumed sex-reversed males had typical testes without intersex (Shimasaki et al. 2003). Zebrafish exposed from 0 to 70 days post-fertilization (dpf) to 0.1 ng/L of TBT showed a male-biased population and produced a high incidence of sperm lacking flagella. At 1 ng/L, the motility of sperm was significantly lower than that of control fish, while at 10 ng/L, all sperm lacked flagella, and at 100 ng/L, milt volume increased. Male sex ratio shifts were similar after exposure from 0 to 70 dpf and 0 to 30 dpf. Equally important, 100 ng/L resulted in 65% males after exposure from 30 to 60 days emphasizing the point that timing of the exposure is very important. Effects on sperm motility and morphology and on milt volume were less pronounced after 30 days than after 70 days of exposure (McAllister and Kime 2003). From this study, the NOEC and LOEC values were 0.01 and 0.1 ng/L, based on nominal exposure concentrations. In a recent study, Lima et al. (2015) exposed zebrafish larvae from 5 dpf up to 120 dpf to 1466 ng TBT/g diet. Animals were fed this diet three times per day. Females were significantly smaller and weighed less, while no change in male total length or weight was observed. Gonad weight in males was significantly heavier but no change was observed in females. There was a 62% decrease in fecundity but no changes in egg viability or hatchability. Overall, sex ratios shifted towards females in contrast to other studies reported above. This could be a strain difference or a concentration effect. The expression of gonadal aromatase was unaffected but in female brain, TBT down-regulated CYP19a1b mRNA. There was also a brain-specific down-regulation of $PPAR_{\gamma}$ in both males and females. TBT effects in zebrafish may involve modulation of $PPAR_{\gamma}$/RXR and brain aromatase based on this study. A LOEC for all endpoints of 1466 ng TBT/g diet was reported, based on measured concentrations.

Reduced sperm counts were observed in guppies (*P. reticulata*) following exposure to 11.2 or 22.3 ng TBT/L for 21 days, with possible effect on Sertoli cell function. This potentially occurred via apoptosis, which could block the nutritional activity of Sertoli cells on maturing spermatids and thereby arrest the release of gametes. TBT exposure for 21 days decreased sperm counts in guppies by 40–75% but flagellar length was unaffected. However, this exposure involved adult fish and effects on sperm were short-term. Sperm counts declined approximately 62–69% but there was no change in testes size or sperm length (Haubruge et al. 2000). In a histological evaluation of TBT's toxicity on spermatogenesis, Mochida et al. (2007) exposed mummichog (*Fundulus heteroclitus*) to mean measured TBT concentrations of 0 (control and solvent control), 0.20, 0.54, 1.0, 1.7, 1.9, and 2.8 μg/L. In this study, there was a relatively small group size and no mention of the methods used to minimize sampling or observational bias. Some of the histopathological changes were difficult to confirm at low magnification, and there were several low quality figure images (changes could also be autolysis). However, damage to epithelial cells of seminal ducts, and slight decrease in spermatozoa numbers were reported with a NOEC-based on histopathology of 1.7 μg TBT/L. TBT can also affect sexual behavior and reproduction in medaka (*Oryzias latipes*) (Nakayama et al. 2004).

Growth effects in fish were also found at very low tissue concentrations. Shimasaki et al. (2003) reported a statistically significant decrease in body weight and length at 18 ng/g body wet weight; however, Meador et al. (2011) found significant increases at essentially the same whole-body concentration in Chinook salmon. Increased growth and lipid content data reported by Meador et al. (2011) are consistent with the mammalian response data characterizing TBT as an obesogen (Grün et al. 2006).

Several studies examined effects of maternally transferred TBT using different routes of exposure including dietary (Nakayama et al. 2005; Shimasaki et al. 2006) and injection (Hano et al. 2007). Adverse effects were noted when TBT concentrations were approximately 5–160 ng/g egg wet weight. These studies support the conclusion that fish embryos are very sensitive to TBT and indicate that maternal transfer is an important route of exposure.

5.4.4 Life-Cycle Studies

Mochida et al. (2010) exposed mummichog *Fundulus heteroclitus* in a fish full-life-cycle assay from the embryo stage until the hatch of the F1 generation at nominal TBTO concentrations of 0 (control and solvent control), 0.13, 0.25, 0.50, and 1.0 μg/L. The mean measured equivalent TBT ion concentrations corresponding to these exposure groups were 0, 0.054 ± 0.005, 0.12 ± 0.02, 0.26 ± 0.02 and 0.37 ± 0.05 μg/L, respectively. In a second experiment, nominal concentrations of

0.13, 0.50 and 2.0 μg TBTO/L were measured as 0.034 ± 0.00, 0.21 ± 0.07 and 0.81 ± 0.02 μg/L, respectively as the equivalent TBT ion. In the F0 generation, TBT exposure resulted in a male-biased sex ratio, an increase in the frequency of the appearance of apoptotic cells in the testis in maturing stages, and a decrease in fecundity. In the F1 generation, time to hatch and hatchability were all markedly affected. Exposure did not affect the proliferation of the germ cells in the testes; however, a significant increase in the number of apoptotic cells in the testes was induced. LOECs for sex differentiation (towards males), reduced spermatogenesis (increased apoptotic cells), and reduced hatching were 0.26, 0.06, and 0.05 μg/L, respectively.

5.5 *Amphibians*

5.5.1 In Vitro Assays

Few studies have been identified that examined the effects of TBT on amphibians in vitro. Using transiently transfected Cos7 cells, Grün et al. (2006) demonstrated that exposure to 60 nM TBT (presumably TBTCl, although compound, purity and source were not specified) activated RXRα and RXRϒ from the amphibian *Xenopus laevis*. Choi et al. (2007) found that TBT inhibited *Rana dybowskii* oocyte maturation in vitro (ED50: 0.6 and 0.7 μM).

5.5.2 Physiological Responses

Mengeling et al. (2016) found that 1 nM TBT (equiv. 290 ng/L) greatly potentiated the effect of T3 on thyroid hormone-induced morphological changes in *X. laevis* but showed both gene and tissue specificity in this capacity. The data also demonstrated that as an RXR agonist, TBT can disrupt TH signaling with outcomes identical to those caused by synthetic RXR-selective ligands and suggested that TBT is not activating a permissive NR-RXR heterodimer such as PPARϒ to achieve this effect.

5.5.3 Organismal Effects

Amphibian embryo development and tadpole metamorphosis are also sensitive to TBT exposure. *Xenopus tropicalis* embryos showed developmental and survival effects when exposed to TBT at 50–400 ng/L in the frog embryo teratogenesis assay-*Xenopus* (FETAX) (Guo et al. 2010). These effects were time- and concentration-dependent, with significant mortality at each time interval (24, 36 and 48 h) and exposure concentration. The most common malformations in the embryos were abnormal eyes and skin hypo-pigmentation, with increased time of exposure; additional common malformations included enlarged proctodaeum and narrow tails fins in tadpoles. Thyroid hormone is linked to eye development in *Xenopus* embryos. The authors suggest that the eye malformations and other malformations are linked to TBT exposure through binding to RXRs and that RXRs form heterodimers with the thyroid hormone receptors.

5.5.4 Life-Cycle Studies

Shi et al. (2014) exposed *X. laevis* to TBT in an Amphibian Metamorphosis Assay (AMA) (OECD TG 231) and a complete AMA (CAMA), which exposed *X. laevis* from Nieuwoop and Faber (NF) stage 46 to stage 66. They found TBT to have anti-thyroid activity in the AMA at TBT concentrations of 12.5–200 ng/L, based on decreased hind limb length in the absence of growth effects (body weight and snout to vent length) or overt toxicity, delayed development by one or two development stages and thyroid lesions characterized by mild increases in thyroid follicle height and/or mild increases in layers of follicular epithelium, and colloid depletion. The CAMA confirmed developmental delays based on front limb emergence and total metamorphosis time; however, these effects were seen in the presence of decreased body length and weight at metamorphosis at 10 and 100 ng/L. The CAMA also found that the intersex and sex ratio increased in favor of males with increasing concentrations of TBT. The intersex gonads had an ovarian cavity with testis-like tissue structure. Apical endpoints in *Xenopus* sp. metamorphosis and embryo eye development were affected by TBT concentrations as low as 12.5 ng/L (Shi et al. 2014). Although these endpoints are regulated by thyroid hormones, it is unclear from this report whether TBT acts directly on thyroid hormones, or indirectly through binding to RXRs.

5.6 Birds

5.6.1 In Vitro Assays

No in vitro studies of TBT effects on birds have been identified.

5.6.2 Physiological Responses and Organismal Effects

Although few studies of the effects of TBT have been conducted in birds, these have demonstrated some reproductive effects from exposure. A subchronic toxicity/reproduction study was performed in Japanese Quail (*Coturnix japonica*) fed a diet containing 0, 24, 60, and 150 mg TBTO per kg diet for 6 weeks (Coenen et al. 1992). No overt toxicity or treatment-related pathological or histological abnormalities were noted in parent birds, and there were no significant effects on egg production, serum alanine aminotransferase (ALAT), serum total thyroxine (TT4), luteinizing hormone (LH), or retinol levels. Limited effects on hematology and no effects on serum biochemistry were found in these birds (Coenen et al. 1994). However, a significant decrease in hatchability and increase in embryo mortality were observed at the two highest doses. Serum calcium values determined throughout the reproduction period were found to be significantly reduced in female birds at all concentrations tested (Coenen et al. 1992). This same study was repeated by five laboratories in an inter-laboratory comparison test (Schlatterer et al. 1993), with the addition of one higher dose group (375 mg/kg feed). Results from this comparison were similar to those reported by Coenen et al. (1992). Dose-related decreases in egg weight, egg production, fertility, hatching success and survival of 14-day-old chicks were observed in most of the laboratories. The NOEC for egg weight and hatchability was 60 mg TBTO/kg feed. Faqi et al. (1999) conducted a follow-up study to examine whether a difference in responses could be observed between a 6-week exposure and a 13-week exposure to 150 and 375 mg TBTO/kg feed. The number of eggs laid, mean egg weight, fertility and hatchability were significantly lower and the percentage of cracked eggs was significantly higher at 375 mg/kg at weeks 6 and 13. Reduced eggshell thickness was observed at week 6. No effects were noted on hematological and clinical chemistry data obtained at weeks 6 or 13 and histological preparations of the organs showed no morphological changes. However, none of these studies in quail examined steroid hormone levels or any related genomic indicators from the HPG axis, nor were any effects on nuclear receptors investigated.

5.6.3 Life-Cycle Studies

No life-cycle studies of TBT effects have been reported for birds.

5.7 *Mammals*

5.7.1 In Vitro Assays

Studies using mammalian in vitro systems have examined interactions between TBT and nuclear receptors and the effects of TBT on enzymes involved in steroidogenesis. Saitoh et al. (2001a) found that TBT had relatively high binding affinity for androgen receptor (AR), with an IC50 of 7.6 μM, but no affinity for estrogen receptor α (ERα). Additional evidence for TBT's impacts on segments of the steroidogenic cascade was presented by McVey and Cooke (2003), who observed decreased 17-hydroxylase and 3β-hydroxysteroid dehydrogenase (3β-HSD) activity in rat testis microsomes at 12 and 59.0 μM TBTCl. Treatment of human granulosa-like tumor cells for more than 48 h with 20 ng/mL TBTCl (~60 nM), significantly suppressed aromatase activity and estradiol production (Saitoh et al. 2001b). Cooke (2002) and Heidrich et al. (2001) reported that TBT is a competitive inhibitor of human aromatase in vitro at 12 and 59 μM TBTCl and 5 and 50 μM TBTCl, respectively, with an IC50 of 6 μM calculated by Heidrich et al. (2001). Together, these studies initially suggested a possible inhibitory role for TBT in steroidogenesis as the primary means of endocrine disruption. However, Kanayama et al. (2005) found that TBT (10, 30, and 100 nM TBTCl) induced the transactivation function of RXR_{α} and $PPAR_{\Upsilon}$ at concentrations lower than those causing aromatase inhibition. The effect of TBT on RXR_{α} is as strong as that of its endogenous ligand, 9-cis-RA and because TBT enhanced protein-protein interaction between RXR_{α} and TIF2, the data suggested that TBT activates transcription via these receptors.

Because TBT induced the transactivation function of RXR_{α} and $PPAR_{\Upsilon}$ at concentrations lower than those required for inhibition of aromatase activity, Kanayama et al. (2005) proposed that this receptor-based pathway is the more likely route for low-dose effects such as imposex in gastropods. The greater likelihood of a role for RXR and PPAR over the aromatase-inhibition hypothesis in TBT-mediated toxicity was further supported by studies that reported opposing effects of TBT on aromatase (CYP19). Sharan et al. (2013) found that low doses of TBT (25, 50, and 100 nM TBTCl) increased CYP19 enzyme activity, mRNA expression, and estradiol production in MCF-7 cells and acted as an ERα agonist. Evidence for the induction of aromatase by TBT was initially provided by Nakanishi et al. (2002) at 30, 100, and 300 nM TBTCl in a human choriocarcinoma cell line. Evidence for the role of the RXR homodimer in aromatase induction in placental tissue was provided by Nakanishi et al. (2002, 2005) at 1, 10, and 100 nM TBTCl. However, the utility of these studies for risk assessment is weak due to the use of inappropriate statistical methods. The variation in results of these in vitro studies with regard to aromatase and ER activation is likely due to wide differences in concentrations of TBT and/or the sources for the cells used in each study, suggesting that TBT's effects may be dose- and tissue-specific.

Additional early evidence for interaction of TBT with RXR and PPAR via direct ligand binding was presented by Grün et al. (2006) using Cos7 cells (transformed Green Monkey kidney fibroblast cells) transfected with human, mouse, and frog (*Xenopus laevis*) nuclear receptors. These experiments showed activation of $RXR_{\alpha-\Upsilon}$ and slightly weaker activation of $PPAR_{\Upsilon}$ and $PPAR_{\delta}$ (same for $PPAR_{\beta}$) at 60 nM TBT (presumably TBTCl, although compound, purity and source were not specified). le Maire et al. (2009) showed that TBT activates all three RXR-PPAR heterodimers (α, β and ϒ) primarily through its interaction with RXR. In contrast to the interaction between TBT and RXR, the active receptor conformation between TBT and $PPAR_{\Upsilon}$ was less efficiently stabilized, making this side of the heterodimer a less efficient binding target for TBT (le Maire et al. 2009), confirming the observations of Grün et al. (2006).

Multiple lines of in vitro evidence exist demonstrating the roles of RXR and PPAR in TBT-induced adipogenesis in mammals. Histological examination of mouse 3T3-L1 preadipocyte cells treated with 100 nM TBTCl by Kanayama et al. (2005) revealed induction and promotion of adipocyte differentiation. This observation was supported by induction of the adipocyte-specific fatty acid-binding protein (aP2) mRNA expression and triglyceride levels in a dose-dependent manner at 10, 30, and 100 nM TBTCl, and linked to the $PPAR_\Upsilon$ pathway by induction of $PPAR_\Upsilon$ mRNA. Induction of adipocyte differentiation was also demonstrated by Grün et al. (2006) using histology and aP2 mRNA expression in mouse 3T3-L1 cells dosed with 10 and 100 nM TBT (presumably TBTCl, although compound, purity and source were not specified). The aP2 promoter contains an RXR:PPAR response element, implicating this pathway in the observed changes. In vitro exposure of mouse multipotent stromal stem cells to 5 and 50 nM TBT (presumably TBTCl although exact compound and purity not specified) increased adipogenesis, cellular lipid content, and expression of adipogenic genes (Fapb4, $PPAR_\Upsilon$, LEP) and decreased mRNA levels of the adipogenesis inhibitor Pref-1 (Kirchner et al. 2010). The adipogenic effects of TBT in this study were blocked by the addition of $PPAR_\Upsilon$ antagonists, suggesting that activation of $PPAR_\Upsilon$ mediates the effect of TBT on adipogenesis. TBT also induced $PPAR_{\Upsilon 2}$ and FABP4 protein expression in bone marrow multipotent mesenchymal stromal cells at concentrations >50 nM TBTCl, resulting in lipid accumulation and terminal adipocyte differentiation (Yanik et al. 2011). Interestingly, Belcher et al. (2014) found that TBT-induced human $PPAR_\Upsilon$ in a Chinese Hamster Ovary (CHO) cell line at 1 nM TBTCl and higher concentrations, but exhibited an "inverted U" dose-response curve, with maximal induction at 100 nM TBTCl. The cause of the loss in functional reporter gene activity was unclear.

Further evidence for the critical role for RXR in mediating effects of TBT comes from in vitro studies of effects on thyroid hormone receptors. Using thyroid hormone-responsive HepG2 cells, Sharan et al. (2014) demonstrated that TBTCl treatment induced a dose-dependent decrease in tri-iodothyronine (T3)-induced thyroid receptor (TR) transactivation and altered the expression of TRβ and its co-regulators including SRC-1 and NCoR. Therefore, TBT acts as an antagonist to TRs and inhibits T3-mediated transcriptional activity. However, TRs can form heterodimers with other nuclear receptors, in particular with the RXR. RXR plays a role in both positive and negative gene regulation through thyroid response elements (LaFlamme et al. 2002). Given the potential for TBT to also activate RXR, and the importance of RXR in the negative transcriptional regulation of genes of the hypothalamo-pituitary axis by T3 (LaFlamme et al. 2002), this suggests that TBT's effects on the thyroid axis may involve multiple nuclear receptor pathways.

5.7.2 Physiological Responses

Grün et al. (2006) dosed 6-week-old male mice for 24 h with 0.3 mg/kg bw TBT (presumably TBTCl, although compound, purity and source were not specified) and examined expression of critical transcriptional mediators of adipogenesis such as RXR_α, $PPAR_\Upsilon$, C/EBP α/β/δ, and sterol regulatory element binding factor 1 (Srebf1) as well as known target genes of RXR_α:$PPAR_\Upsilon$ signaling from liver, epididymal adipose tissue and testis. TBT either had no effect, or weakly repressed RXR_α and $PPAR_\Upsilon$ transcription in liver and decreased RXR_α, $PPAR_\Upsilon$, C/EBPα and

-δ in adipose tissue and testis. C/EBPβ was strongly induced in liver and testis, but more weakly induced in adipose tissue. Proadipogenic transcription factor Srebf1 was also induced in adipose tissue. Fatty acid transport protein (Fatp) mRNA levels were up-regulated two- to three-fold in liver and epididymal adipose tissue but not testis by TBT. Additional up-regulation of genes associated with fatty acid synthesis was also noted and together, these gene expression data confirmed TBT as a potential adipogenic agent in vivo.

Thyroid-related effects of TBT have been reported in vivo in mammals (e.g., Decherf et al. 2010; Sharan et al. 2014). Thyrotropin-releasing hormone (TRH) production is controlled at the transcriptional level by T3 through TRs but also via RXR and PPARγ (LaFlamme et al. 2002). Decherf et al. (2010) exposed Swiss wild-type mice to TBT through lactation after dams were gavaged with a single 40 mg TBTCl/kg dose, and examined effects in pups on hypothalamic expression of genes implicated in metabolism and regulated by T3. They found that TBT dose-dependently increased T3-independent transcription from the TRH promoter (i.e., transcription that is controlled through RXR), but had no effect on T3-dependent repression. However, the effect on T3-independent expression was not observed in pups whose mothers were exposed chronically to 0.5 mg/kg by gavage for 14 days. Additionally, this paper demonstrated that exposure to TBT has a two-pronged effect on transcription from the aromatase and tyrosine hydroxylase promoters—it significantly reduced T3-independent transcription but also abolished T3-dependent regulation, confirming the role of TBT as a T3 antagonist that had been reported in vitro. Although the exposure levels in this study were relatively high, the results lend important insights into the hypothalamic effects of TBT.

Thyroid system effects in mice have been demonstrated at significantly lower exposure levels. Swiss albino male mice exposed to three doses of TBTCl (0.5, 5 and 50 μg/kg/day) for 45 days showed hypothyroidal effects (Sharan et al. 2014). TBT exposure markedly decreased serum thyroid hormone levels, which correlated with down-regulation of thyroid peroxidase (TPO) and thyroglobulin (Tg) genes in the thyroid gland and augmented circulating thyroid stimulating hormone (TSH) levels and TSH receptor (TSHr) gene in the thyroid gland. In addition, Pax8, a thyroid-specific transcription factor (mRNA and protein) and sodium-iodide symporter (Slc5a5) (mRNA) were also down-regulated. Sharan et al. (2014) concluded that TBT induces hypothyroidism by suppressing transcriptional activity of thyroid-responsive genes and inhibiting T3 binding to thyroid receptors, thereby preventing recruitment of co-activators and corepressors on the promoters of target genes.

Zuo et al. (2014) exposed male KM mice for 45 or 60 days to 0.5, 5, and 50 μg/kg TBTCl orally administered by gavage once every 3 days, and examined effects on the pancreas, glucose homeostasis, and circulating steroid and thyroid hormone levels. Animals treated with TBT for 60 days exhibited elevated fasting plasma glucose levels and decreased serum insulin and glucagon. TBT treatment for 45 days resulted in a dose-dependent increase in testosterone levels and a decrease in 17β-estradiol levels in the testes and serum compared to the control. Serum T4 levels did not show significant alteration in the TBT-exposed group, while T3 levels showed a reduction in the TBT-exposed group and severe damage of the thyroid gland was observed histologically in mice exposed to 50 μg/kg TBT. No histological damage was observed in the pancreas after TBT exposure for 45 days. However, the number of apoptotic cells in the pancreas increased significantly with dose. TBT treatment for 45 days resulted in a dose-dependent decrease in pancreatic ERα expression but not ERβ levels, and resulted in an elevation of AR expression. This study is the first to examine direct endocrine effects of TBT on the pancreas.

Kirchner et al. (2010) investigated effects of in utero exposure of mice to TBT on adipose-derived stromal stem cells (ADSCs). Pregnant dams received a single 0.1 mg/kg body weight dose of TBT (presumably TBTCl although exact compound and purity not specified) by gavage and stromal cells were isolated from white adipose tissue (WAT) of their 8-week-old pups. Cells from TBT-exposed mice showed increased adipogenic capacity and lipid accumulation, reduced osteogenic capacity, increased Fapb4 and PPAR$_\Upsilon$ mRNA expression, decreased adipogenesis inhibitor Pref-1 mRNA, hypomethylation of the promoter/enhancer region of the Fapb4 locus, and an increased number of preadipocytes in the cells. This study provided the first evidence that in utero exposure to TBT counteracts osteogenesis and induces preferential differentiation of ADSCs into adipocytes.

5.7.3 Organismal Effects

In mammals, in vivo exposure to TBT has been shown to cause reproductive and other apical effects, although early studies used relatively high exposure levels, at which the effects of TBT are unlikely to be through the RXR/PPAR pathway. Harazono et al. (1996) reported a higher rate of pregnancy failure in Wistar rats exposed to 12.2 and 16.3 mg TBTCl/kg body weight. In a follow-up study, female Wistar rats exposed to 8.1, 16.3, or 32.5 mg TBTCl/kg body weight (25, 50, or 100 μM/kg) on days 0 through 3 of pregnancy, or 8.1, 16.3, 32.5, or 65.1 mg/kg (25, 50, 100, or 200 μM/kg) on days 4 through 7 of pregnancy by gastric intubation, and their fetuses, exhibited significantly lower body weights at 16.3 and 32.5 mg/kg than controls (Harazono et al. 1998). Exposure to 16.3 mg/kg and higher produced a significant increase in the rate of implantation failure, and dosing at the same levels on days 4–7 of pregnancy caused a significant increase in the incidence of post-implantation loss (Harazono et al. 1998). The authors concluded that susceptibility to, and manifestation of, the antifertility effects of TBTCl vary with the gestational stage at the time of administration. Dosing of rat dams with TBT beginning on gestational day 8 by oral gavage caused a significant reduction of dam's body weight at 10 mg/kg body weight during gestation and postnatally (Cooke et al. 2008). At postnatal days 6 and 12, neonatal pup weights were reduced at this concentration. However, Cooke et al. (2008) also noted that at the lowest dose of 0.25 mg TBTCl/kg body weight, dam's body weight increased relative to controls. Similarly, Zuo et al. (2009) showed that exposure of male mice to TBTCl at 5 μg/kg body weight for 45 days resulted in an increase in body weight and hepatic steatosis accompanied by hyperinsulinemia, hyperleptinemia, and changes in several metabolism-related hormones. The variation in body weight responses appears to reflect both life stages during exposure and dose, with exposure in utero likely predisposing the animal to increased adipose mass as it ages and high dose exposures resulting in body weight loses.

One of the first studies to examine the effects of in utero exposure to TBT on lipid homeostasis and adipogenesis was conducted by Grün et al. (2006) using pups from pregnant C57BL/6 mice, which were injected intraperitoneally daily from gestational day 12–18 with 0.05 or 0.5 mg/kg body weight TBT (presumably TBTCl, although compound, purity, and source were not specified). Histological examination demonstrated that TBT exposure caused a disorganization of hepatic and gonadal architecture in the pups at birth, and liver sections exhibited signs of steatosis. Adipose mass in 10-week-old TBT-treated males was significantly higher than in controls although no overt increases in body mass were noted.

Effects of in utero exposure to TBT on fetal gonad morphology have been reported in Sprague–Dawley rats (Kishta et al. 2007). Light microscopic evaluation found that the number of Sertoli cells and gonocytes was reduced in fetuses whose mothers were gavaged daily from days 0 to 19 or 8 to 19 of gestation with 20 mg TBTCl/kg. Likewise, large intracellular spaces between Sertoli cells and gonocytes and increased abundance of lipid droplets in the Sertoli cells were observed. Electron microscopy studies revealed abnormally dilated endoplasmic reticulum in Sertoli cells and gonocytes. In the ovaries, TBT (20 mg/kg, days 0–19; 10 mg/kg, days 8–19) reduced the number of germ cells by 44% and 46%, respectively. Kishta et al. (2007) also examined gonadal gene expression in the fetuses and found significant up-regulation of testicular genes related to stress response but no up-regulation of these genes in the ovary. In ovaries, down-regulation was noted of genes involved with signal transduction.

TBT resulted in early puberty and impaired estrous cyclicity in female mice exposed perinatally (1, 10, or 100 μg TBTCl/kg body weight/day from day 6 of pregnancy), although no effects on circulating sex steroids (E2 or T) were observed (Si et al. 2012). Reductions in body weight were also reported by Si et al. (2012). Identical exposures of pregnant mice to TBT dramatically decreased sperm counts and motility in male offspring but had limited effects on intratesticular and serum hormone levels, suggesting that altered expression of receptors rather than hormone levels may be involved (Si et al. 2013).

5.7.4 Life-Cycle Studies

A two-generation reproductive toxicity study was conducted in rats using dietary exposure to TBT to evaluate its effect on sexual development and the reproductive system (Ogata et al. 2001; Omura et al. 2001). Pregnant female rats were exposed throughout pregnancy until weaning to 5, 25, or 125 μg TBTCl/g diet [assuming adult female rats weigh ~150 g and eat 16 g of food per day (US EPA 1988), this equates to approximately 12, 60, and 300 μg/kg body weight/day]. F0 and F1 progeny were provided with the same TBTCl diet as their mothers. For males

(Omura et al. 2001), significant effects on monitored endpoints (reduced body weight, delayed eye opening, reduced testis, epididymis and ventral prostate weights, and decreased spermatid count) were observed primarily at 125 μg/g. Only minimal histological changes were observed in the testes. A dose-dependent increase in serum testosterone occurred only in the F1 rats, and serum E2 was affected only in the 125 μg/g groups of F1 and F2. As is commonly seen in high dose exposures, the data suggest that these results were primarily related to direct toxic effects of TBT rather than functioning through specific endocrine pathways. A similar conclusion was drawn from the results of the companion study on the female offspring of these rats (Ogata et al. 2001). Reproductive outcomes of dams (number and body weight of pups and percentage of live pups) and the growth of female pups (day of eye opening and body weight gain) were significantly decreased in the group exposed to 125 μg TBTCl/g diet.

Chamorro-García et al. (2013) conducted a follow-up study to the one by Kirchner et al. (2010), in which they exposed female C57BL/6J mice prior to conception and during pregnancy to 5.42, 54.2, or 542 nM TBT (presumably TBTCl although exact compound, source and purity not specified) in drinking water (equivalent to 0.53, 5.3, and 53 μg/kg/day) to determine whether prenatal exposure would affect subsequent generations (F1 = exposed in utero, F2 = exposed as germ cells, F3 = no exposure). Prenatal TBT exposure elicited striking transgenerational effects in males including increased white adipose tissue (WAT) depot weights, adipocyte size, and adipocyte number at most doses in the three generations. More modest changes were observed in females, yet most doses of TBT led to significant increases in WAT depot weight and adipocyte size in F1 and F2 animals. Effects on body weight were modest and not directionally consistent in both sexes. Quantitative PCR analysis of adipogenic markers in bone marrow-derived multipotent mesenchymal stem cells (MSC) revealed sharply increased expression of Zfp423 and Fabp4 and decreased expression of Pref-1, an inhibitor of adipocyte differentiation, in TBT males from all three generations. Results from female mice were similar, but with less pronounced changes in F2 and greater variability in F3 mice than in males. Osteogenic markers, ALP and Runx2, sharply decreased in F1 and F3 males and females but were primarily unchanged in F2. All three generations exhibited hepatic lipid accumulation and up-regulation of hepatic genes involved in lipid storage/transport, lipogenesis, and lipolysis. The Chamorro-García et al. (2013) results show that early-life exposure to TBT can have transgenerational effects on adipogenesis, at least through the F3 generation.

6 Adverse Outcome Pathway

The adverse outcome pathway (AOP) is a framework that summarizes existing information for a given biological pathway from the MIE, through various levels of biological organization (genetic, molecular, physiological, and organismal), and culminates with population-relevant results that can be used for ecological risk assessment. This framework was described by Ankley et al. (2010) and has been utilized many times to highlight progressive linkages between receptor activation and population-relevant outcomes.

The present review provides the necessary information needed to populate the various AOP components, which are displayed in Fig. 3. It should be noted that this is a putative and most likely incomplete AOP that was drafted based on existing information and our current ability to interpret the available data. In order to obtain the complete AOP, far more work is needed to fill in the gaps between MIEs and key events (KE) across the different taxa and life stages (applicability domains); even more work is required to obtain a quantitative AOP as it is likely that different pathways are operating and their LOECs may be different between applicability domains. Since TBT can trigger more than one AOP, the potential for an operational AOP network is substantial and elucidation on whether this network is converging, diverging or independent in terms of adversity will be needed. All the information captured in Fig. 3 originates from invertebrate (primarily molluscan) studies, although certain pathways have been confirmed in other taxa. Besides the putative and incomplete nature of the AOP presented here, it is almost certain that the critical pathway appears to be initiated by the RXR and RXR/PPAR interactions, to include imposex as a key event and to result in complete reproductive failure and decline population trajectory.

The full AOP can only be completed when some basic aspects of endocrinology of invertebrate species, that are currently largely unknown, become available. The same holds true for the RXR and RXR/PPAR modulated pathways in vertebrate species, although some information has recently become available (Tonk et al. 2015). Nevertheless, it should be noted that unlike the majority of the currently available AOPs in the context of chemical perturbations, the adverse outcome of TBT exposure at population level is well established at least for molluscan species as the numerous publications on population extinctions globally testify.

7 Species Sensitivity Distribution (SSD) for Toxic Effects in Aquatic Organisms

7.1 *Water Exposure SSDs*

Species sensitivity distributions (SSDs) have been derived using apical, population-relevant data (reproductive outcomes, including fertilization, embryonic development, hatching, larval and juvenile growth, as well as sex ratio) mainly from long-term aqueous exposure studies matching Levels 4 and 5 in the OECD CFEDTA (see Tables 2 and 3 of the Annexes). These SSDs do not necessarily include all relevant species and endpoints and are not expected to be comprehensive. However, they have been built using robust endpoints from reliable studies, and therefore accurately reflect the sensitivity of major groups of aquatic organisms. Two SSDs have been constructed—one using LOECs (Fig. 4) and the other NOECs (Fig. 5) from chronic studies—using the ETX 2.1 software of the Netherlands National Institute for Public Health and the Environment (RIVM) (van Vlaardingen et al. 2014).

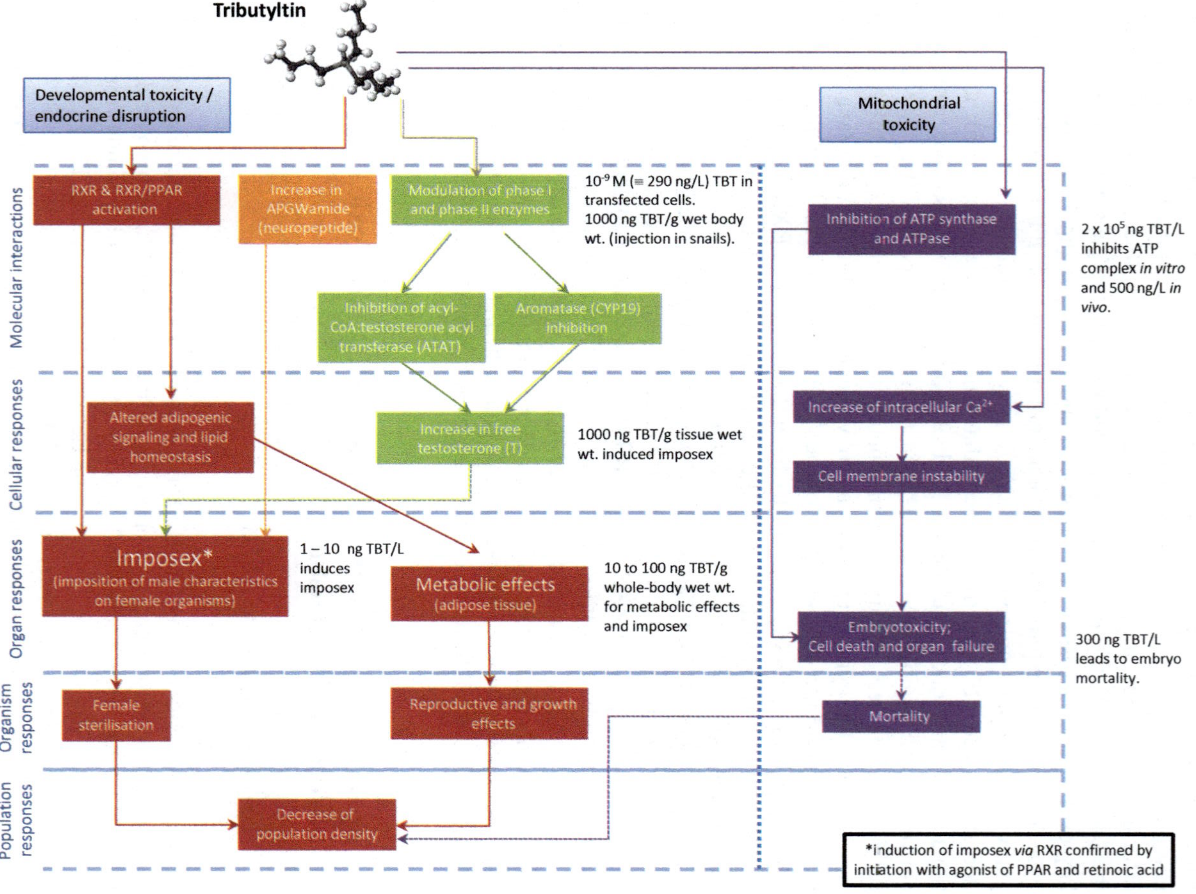

Fig. 3 Proposed adverse outcome pathway (AOP) for TBT toxicity

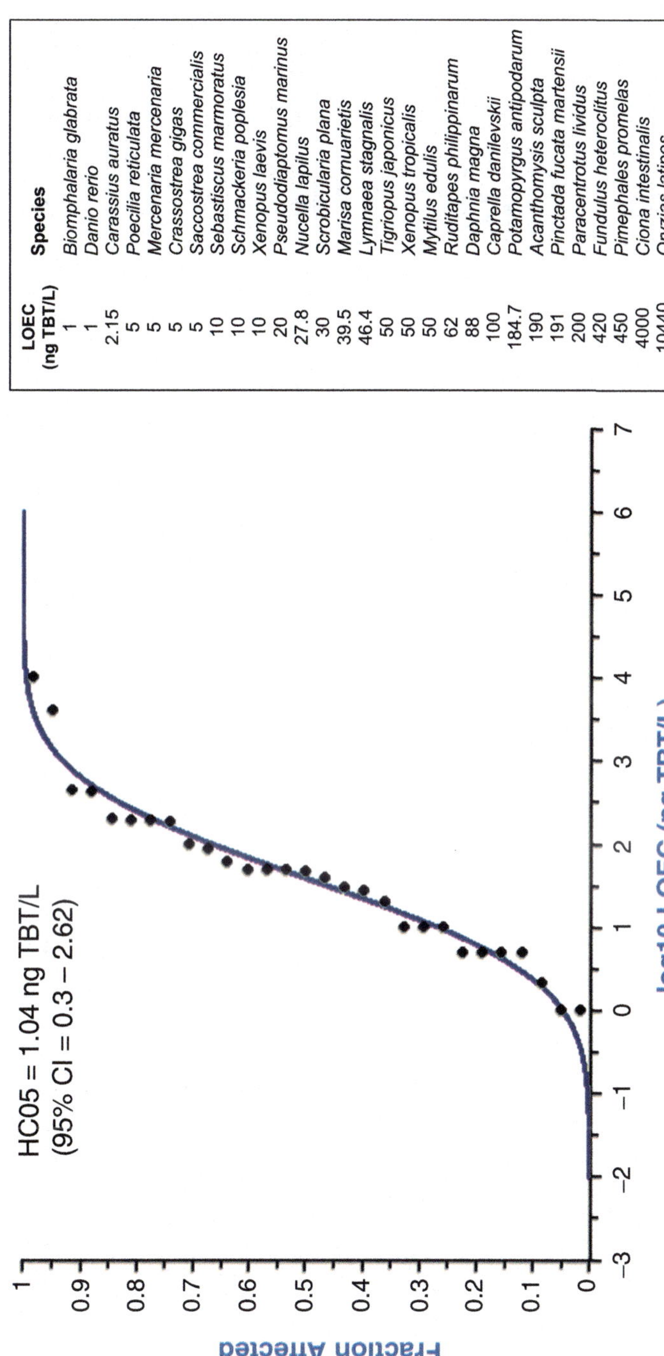

Fig. 4 Species sensitivity distribution (SSD) constructed with sublethal population-relevant LOEC values for TBT in 29 aquatic species. See Table 2 of the Annexes for the endpoint value and description for each species

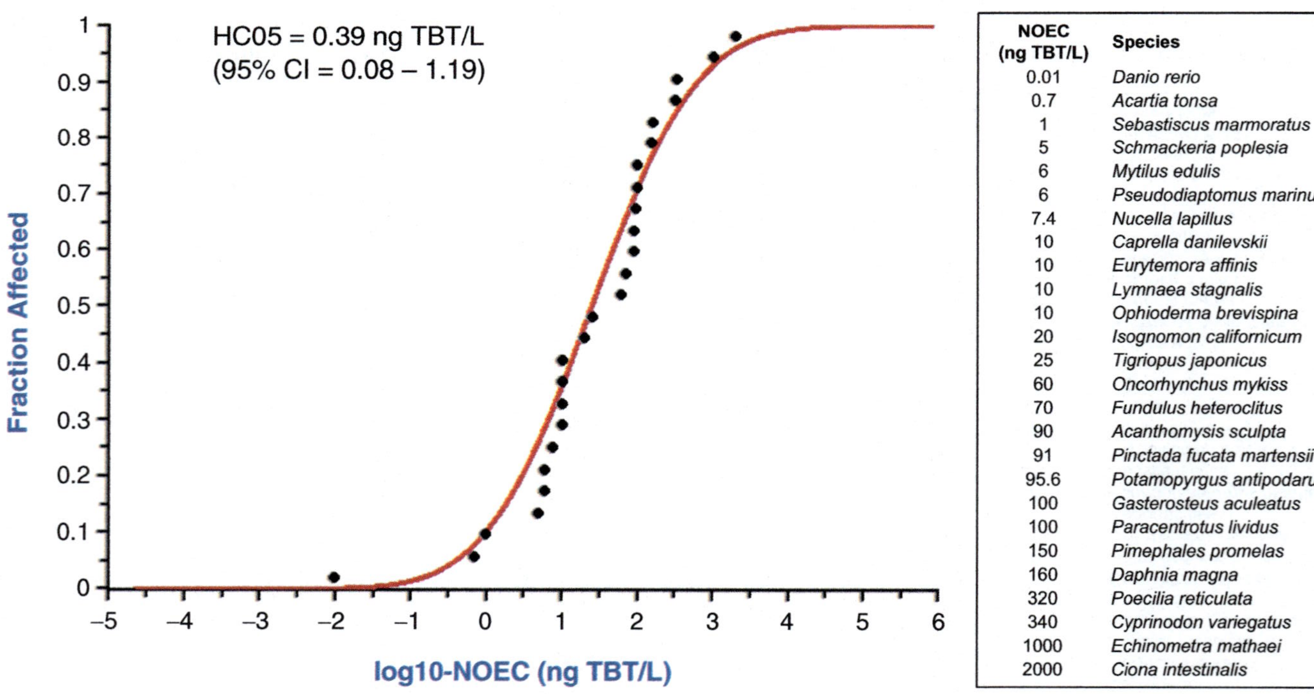

Fig. 5 Species sensitivity distribution (SSD) constructed with population-relevant NOEC values for TBT in 26 aquatic species. See Table 3 of the Annexes for the endpoint value and description for each species

When several biological endpoints were available for one given species, only the lowest LOEC and/or NOEC was used, to ensure that each species appeared only once in the SSD. In some cases, the LOECs are unbounded, so the true effect concentration could be much lower. The NOEC is also sometimes significantly lower than the LOEC, depending on test design (i.e., some may be over-precautionary). By assessing the distribution of both LOECs and NOECs, these factors have been ignored for the purposes of this analysis. For both SSDs, the goodness-of-fit was acceptable (significance level: 0.01–0.1), as assessed using the Anderson-Darling, Kolmogorov-Smirnov, and Cramer von Mises tests for normality.

As a result from the NOEC-based SSD, 0.39 ng/L (95%-CI: 0.08–1.19 ng/L) was determined to be the HC5 (the predicted concentration that affects 5% of the species in the distribution), whereas the HC5 based on LOEC values was 1.04 ng/L (95%-CI: 0.3–2.62 ng/L). The NOEC-based HC5 value is close to current EU regulatory thresholds (e.g., 0.2 ng/L; EU 2005), which incorporates a safety factor and so should be treated with caution. For comparison, the USEPA ambient water quality criterion (seawater chronic) has been set to 7 ng/L (US EPA 2003).

An interesting feature of this sensitivity analysis is that some fish and other invertebrate species are more sensitive to TBT than caenogastropod molluscs, in particular some copepods (*Acartia tonsa*, *Schmackeria poplesia Pseudodiaptomus marinus*) and the fish *Danio rerio* and *Sebastiscus marmoratus*. The lowest endocrine-sensitive LOEC identified in this review was 0.1 ng TBT/L for a male-biased sex ratio change and abnormal sperm in zebrafish (McAllister and Kime 2003). Our analysis found that study convincing, well designed, and described. Aqueous concentrations of TBT in aquaria for the highest dose (100 ± 5 ng TBT/L) were verified by gas flame atomic absorption spectrophotometry while other doses were below the detection limits. Because this is the lowest value, it sets the lower limit concentration for the TBT risk assessment for fish and also for molluscs and other invertebrate species.

Such a high sensitivity of zebrafish sex ratio to TBT is consistent with the fact that TBT inhibits aromatase, thus increasing the level of free testosterone to which zebrafish appear to be extremely sensitive (Holbech et al. 2006; Örn et al. 2006; OECD 2012b). There is therefore a biological plausibility that links the molecular initiating event (aromatase inhibition) to the key event (increased levels of free testosterone) subsequently leading to an apical effect (male-biased sex ratio), which is meaningful for inferring potential impact at the population level. Such causality relationships were thoroughly described by Matthiessen and Weltje (2015) for azoles compounds, which also act as aromatase inhibitors.

Comparison of the potencies of known aromatase inhibitors, such as prochloraz, with TBT requires an in-depth analysis of the available data. However, this is somehow out of the scope of this review. Nevertheless, it appears that, with an IC_{50} around 0.2 μM (Cooke 2002), TBT is a weaker aromatase inhibitor in vitro as compared to prochloraz ($IC_{50} = 0.04$ μM; Vinggaard et al. 2000). In contrast, in vivo (zebrafish fish sexual development test), a dramatic shift in potencies was observed with TBT affecting sex ratio towards males at 1 ng/L (McAllister and

Kime 2003) whilst prochloraz shifted sex ratio towards males at 202 μg/L (Kinnberg et al. 2007). It is highly unlikely such a large difference stems from slightly different exposure conditions. Hence, it could be speculated that TBT exerts its actions on vertebrate sexual development via additional mechanisms that enhance the effects of aromatase inhibition alone. On the other hand, direct comparison is hindered by the fact that prochloraz along with other pesticides have additional endocrine modulating properties (i.e., androgen and estrogen receptor interactions). A thorough review of these data is required before firm conclusions can be made on the primary adverse outcome pathway of TBT in vertebrates.

There were various effects (including in vitro changes in genomic markers and enzyme activities) in other fish species within at least one order of magnitude of the results obtained by McAllister and Kime (2003) but these endpoints and LOEC values for zebrafish stand out as the most ecotoxicologically relevant. This suggests that the endpoints used for other fish studies (often growth) were not the most relevant. However, the dramatic effect observed in zebrafish may be restricted to fish species that share the same pattern for sexual differentiation rather than all fish. Zebrafish gonads initially develop as ovaries, however in male fish, the ovarian tissue degenerates and the testis develop (Maack and Segner 2003). This period of juvenile hermaphroditism (Takahashi 1977) may explain the increased sensitivity of the sex ratio endpoint after exposure to AR agonists and aromatase inhibitors during critical developmental windows.

7.2 *Tissue Residue-Based Analysis*

A given ambient toxicity metric (e.g., EC_{50} or LOEC) that is based on aqueous or sediment concentrations can result in a range spanning orders of magnitude for different species. The equivalent tissue residue toxicity metric (e.g., ER_{50} or LOER) often exhibits lower variability (Meador 1997). This has been observed for a large variety of taxa and chemicals, which has been discussed in many publications (Meador 2006; Meador et al. 2008; McElroy et al. 2011) and is known as the tissue-residue approach for toxicity assessment. TBT provided one of the first examples of the utility for this approach based on a large database of tissue residue toxicity data for mortality and reduced growth. The analysis of TBT toxicity has expanded from evaluating these high dose endpoints to endocrine-related responses in invertebrates and fish responding to very low environmental concentrations. While some of the datasets are limited, they do indicate a relatively consistent response among species for a given endpoint and whole-body tissue concentration (Meador 2011). Widely variable toxicokinetics among species is the main factor responsible for the high interspecies variability for a given toxicity metric based on external dose. Consequently, when internal dose is used to determine a toxicity value, toxicokinetic differences are not a confounding factor, which results in greatly reduced variability among species response values (Meador et al. 2008; McCarty et al. 2011).

An analysis of the imposex endpoint as a function of whole-body tissue concentrations is shown in Fig. 6. The program SSDMaster (Rodney and Moore 2008) was

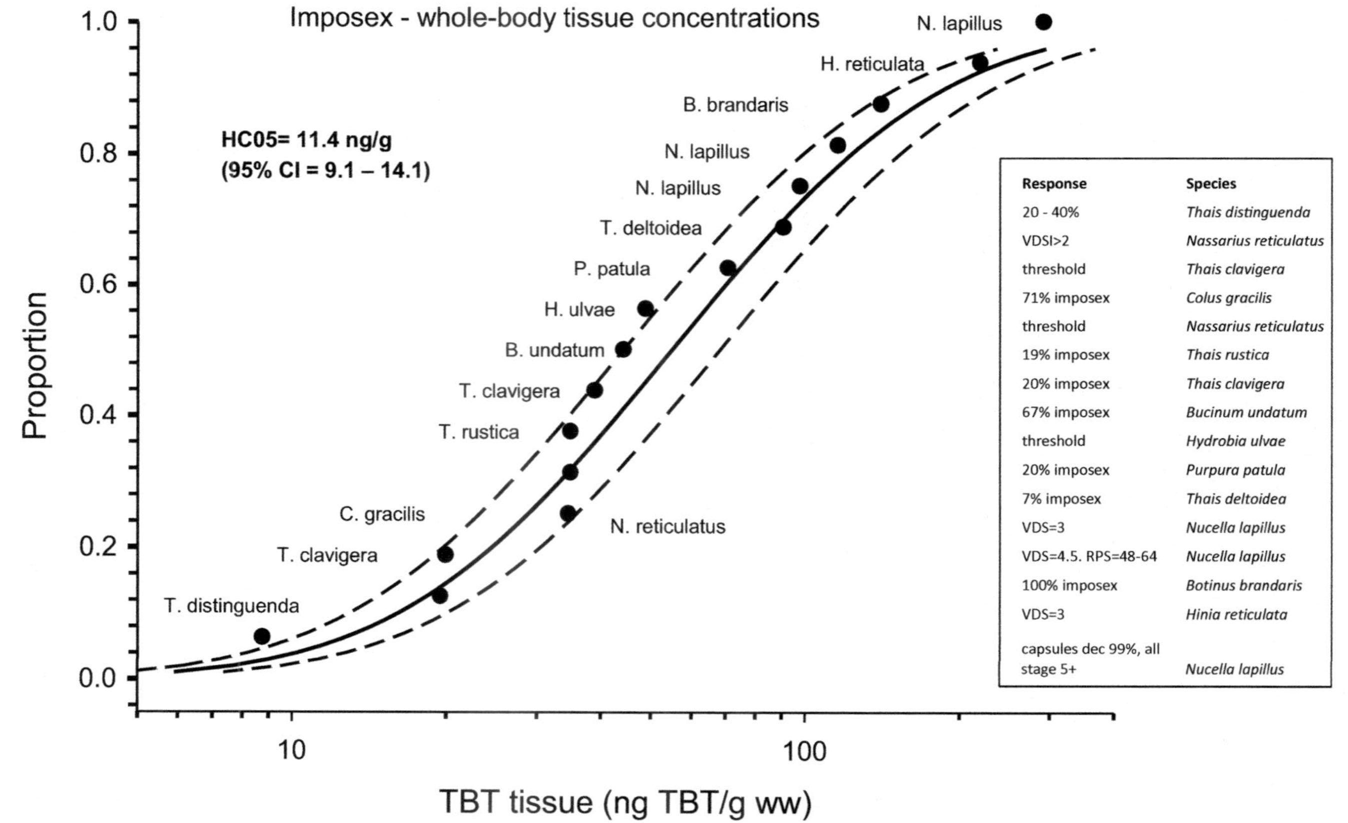

Fig. 6 Cumulative distribution function (CDF) constructed with tissue TBT concentrations (as ng TBT/g wet weight) that induce imposex in Caenogastropod molluscs

used to generate the data for this plot. The cumulative distribution function (CDF) shows that the sensitivity of one given species may vary according to the imposex stage and TBT tissue burden. These data indicate a relatively narrow range of concentrations for the imposex response spanning from threshold to 100% induced. Based on these data, the hazard concentration resulting in low level effects was determined to be 11.4 ng TBT/g wet weight (95% CI = 9.1–14.1). This concentration characterizes a low level response; hence, a safety factor may be needed to determine the potential "no effect" level. The tissue concentrations in this CDF have utility in assessing population fitness for these gastropods in the field.

8 Sources of Uncertainty, Data Gaps, and Confounding Issues

8.1 Transgenerational Effects

TBT is highly bioaccumulative and maternal transfer to eggs has been demonstrated (e.g., Inoue et al. 2006; Ohji et al. 2006). Effects can occur over multiple generations in some invertebrate species, e.g. chironomids (Lilley et al. 2012) and copepods (Huang et al. 2006), as well as in fish, birds, and mammals. However, it is not clear whether or not this is exclusively linked to endocrine-mediated mechanisms.

8.2 Sensitive Species

The general perception is that molluscs are the most sensitive taxonomic group to TBT exposure due to the observation of imposex in wild species at low environmental concentrations. However, the present analysis determined that a number of fish species and other invertebrate species (e.g., copepods) have similar or greater levels of sensitivity when considered in terms of population-relevant responses.

8.3 Potency

With respect to endocrine-mediated effects, TBT is highly potent as it can act in aquatic organisms at levels of parts per trillion (ng/L) and lower (Figs. 4 and 5), and low ng/g tissue concentrations (Fig. 6), whereas lethal toxicity occurs at much higher concentrations. Thus, the HC5 value derived from an SSD based on LOECs for mortality (Fig. 7) is ten times higher than the HC5 estimated from LOEC for sublethal effects (Fig. 4). In the copepod *Acartia tonsa*, the NOEC values for

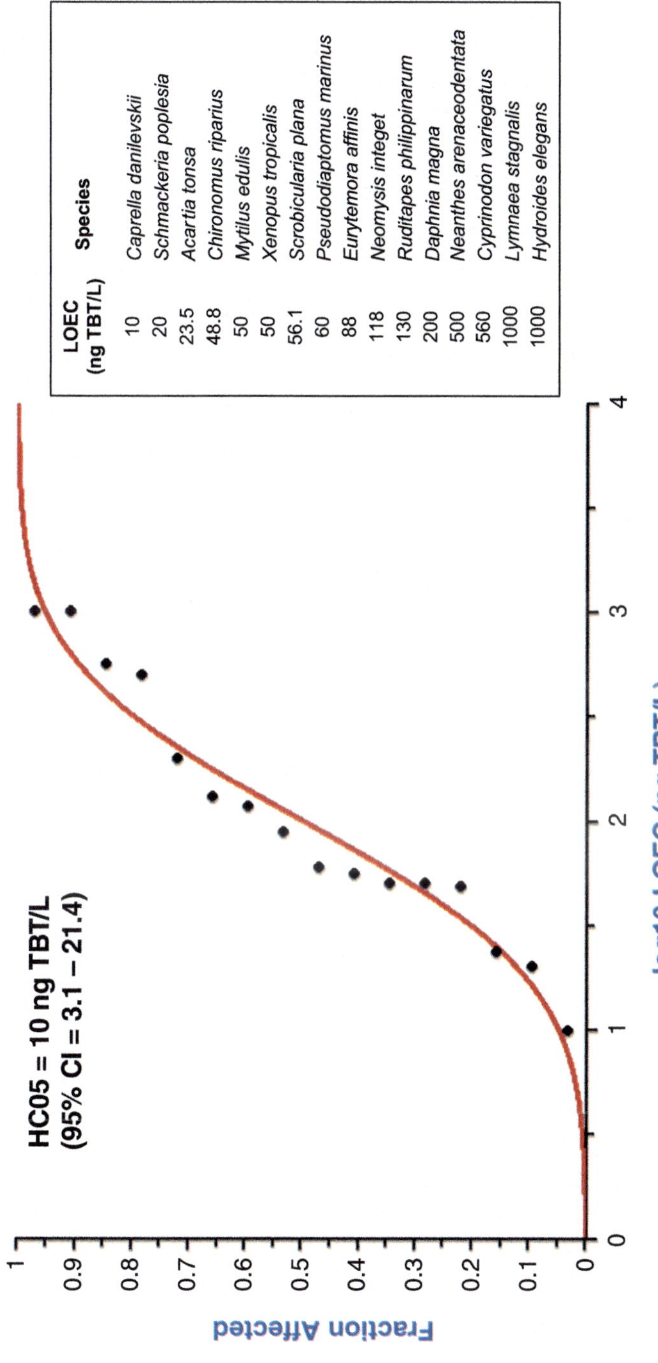

Fig. 7 Species sensitivity distribution (SSD) constructed with TBT LOEC values for mortality in 16 most sensitive aquatic species. Data that were used to construct this SSD are given in Table 4 of the Annexes). The goodness-of-fit was acceptable (significance level: 0.01–0.1), as assessed using the Anderson-Darling, Kolmogorov-Smirnov, and Cramer von Mises tests for normality

reproductive effects and mortality are 0.7 and 11 ng/L, respectively (Kusk and Petersen 1997).

Potency of TBT is also shown at the molecular level where molecular initiating events are elicited at concentrations that are several hundred times lower than those causing basal toxicity (Fig. 3).

8.4 Non-monotonic Dose-Response or Lack of a Threshold Dose

Inverted U-shape responses to TBT have been observed for some endpoints, depending on the concentration, mainly involving gene expression studies (e.g., Mortensen and Arukwe 2007; Kortner et al. 2010; Morales et al. 2013; Pascoal et al. 2013). From in vivo studies, there is also some evidence that TBT has been shown to display an inverted U-shape response for several endpoints. A good example is the impact of TBT on body weight in fish and mammals (Cooke et al. 2008; Meador et al. 2011; Si et al. 2011). Meador et al. (2011), for example, found that TBT exposure in fish enhanced growth and lipogenesis at low doses and inhibited growth and reduced lipid content at high doses; this was attributed to two modes of action operating at different doses. Most likely, the cause of a non-linear response for TBT is different dose-dependent mechanisms of action. TBT is known to be an uncoupler of oxidative phosphorylation and directly affects ATPase (ATP-synthase). This enzyme that plays an important role for providing cellular energy is located within the mitochondrial membrane and consists of two regions: the Fo section, embedded within the membrane and the F_1 section, outside the membrane but inside the matrix of the mitochondria. TBT interacts with the Fo section of ATPase. When exposures occur at higher doses, this is the most likely MeOA that causes reduced growth and death. Baseline toxicity (narcosis) is not possible because all species die at about 10^4 ng/g whole-body concentration, far below baseline doses. Hence, the low dose effects are definitely due to RXR-PPAR binding but high dose responses are due to mitochondrial dysfunction. Of course, the U-shaped response will depend on the endpoint; it is highly plausible for metabolic effects. The question is whether this non-linear behavior will occur for other responses. Such non-linear concentration-response should be considered with caution regarding the magnitude and consistency of the changes. It is important to determine whether or not these responses are significant, and if they consistently occur in different species. One should also consider whether U-shaped dose-responses are a result of endocrine-mediated perturbations, multiple mechanisms of action, or adaptive process (see Parrott et al. 2017, and the references therein). They also may relate to exposure artifacts (e.g., exceeding chemical solubility, erroneous doses, or failure of chemical delivery systems).

9 Areas for Future Research

Although TBT has now been recognized as an EDC at a global scale, if its ED properties were originally assessed at low concentrations using a limited group of USEPA Tier-1 tests that did not include the AMA, further testing may not have been triggered. The current Tier-1 tests cover primarily three hormonal pathways (EAT) and as such do not target the primary endocrine mode of action of TBT, via RXR/PPAR activation. Based on the literature reviewed here, TBT would elicit positive results in the steroidogenesis and aromatase assays only at levels much higher than those needed to elicit an endocrine effect via the RXR/PPAR interactions. The lack of testing for other hormonally and metabolically important nuclear receptors such as PPAR and RXR under both the US and OECD testing batteries has been recognized as a deficiency and prioritized for assay development by OECD (OECD 2012a). The results of the present review underscore this need. However, we recognize that assays for RXR/PPAR activity may be more difficult to develop and validate than those for other nuclear receptors such as ER and AR, due to the complex nature of their activation, the formation of homo- and heterodimers with other nuclear receptors, and the permissive/non-permissive nature of downstream effects. In other words, a different platform than the current transactivation assays (e.g., one that allows assessing cross-talk between these receptors) may be needed for a full understanding of perturbations stemming from chemical interactions with these receptors.

In addition to new low-tier testing that goes beyond EAT, suitable higher tier tests should also be developed in order to assess the plethora of biological effects that involve targets other than EAT. To this end, the new OECD Test Guidelines (TG242 and TG243) on molluscan reproductive toxicity are an important development, although such tests do not discriminate between ED and non-ED mediated mechanisms. However, standardized partial or full life-cycle tests should also be developed for additional invertebrate phyla such as annelids and echinoderms as they are not only numerous, but also ecologically important taxonomic groups with a largely unknown endocrine system.

When designing a testing strategy for chemicals that are suspected to act via the endocrine system, ecotoxicologists should incorporate the lessons learned from TBT, as they are both numerous and important. The first relates to apparent species sensitivity. Previously, molluscs were assumed to be the most sensitive group to TBT, primarily due to overt population effects. However, our analyses demonstrated that certain fish species are equally if not more sensitive to its effects. Likewise, the body burden of a chemical should be taken into account especially when the apparent sensitive species are filter feeders. In many cases, these species exhibit elevated bioaccumulation because of higher rates of uptake and lower rates of metabolism compared to other species. Tissue residue toxicity metrics also vastly improve the characterization of toxic responses because the inherent variability in toxicokinetics found among species is incorporated and accounted for.

Another important aspect relates to species extrapolation, particularly when fundamental aspects of endocrine control are largely unknown. TBT is not the only chemical with observed effects in invertebrate species that were subject to erroneous interpretations based on vertebrate endocrinology. The same assumption was made for the role of estrogens on bivalve vitellogenesis (Gagné et al. 2001), which resulted in numerous research programs globally attempting to use bivalve molluscs as model species for studying effects of vertebrate steroids (Scott 2012, 2013). The presence of an estrogen-like receptor in their genome cemented this assumption without any functional characterization of this receptor (Kishida et al. 2005). Only recently this assumption was proven incorrect (Morthorst et al. 2014), highlighting yet again the importance of fundamental knowledge before a sound testing strategy is in place.

Finally, another important lesson learned from TBT is that care should be taken when dealing with chemicals that display multiple mechanisms of action; the actual number of these is unknown but they do exist and can lead to incorrect interpretations of experimental and field data. Retrospective analysis of the TBT data clearly indicates a dual mode of toxicity (low versus high doses) that is a function of different MeOA. This highlights the need for comprehensive testing at different levels and using different species prior to interpretation of data.

10 Conclusions and Recommendations

Environmental TBT concentrations measured post-2008 show that exposure may still occur in the range 0.1–8 ng/L (mean = 0.3 ng/L) in representative European surface waters (UK Environment Agency, pers. com. 2016) and in other locations (e.g., Kim et al. 2014; Ho et al. 2016), suggesting that there is still an environmental risk from legacy contamination. Risks may be higher in regions of the world with less effective enforcement. TBT was introduced on the market in the early 1960s, at a time when regulatory assessment of chemicals was at its infancy. Using deliberate "retrospective thinking," and considering the information gathered in this case study, one important question arises: would TBT be identified as an ED using current screening and testing methods?

Typical endocrine responses are elicited by compounds that mimic estrogen, androgen, and thyroid pathway hormones and act via various nuclear receptors. Tributyltin is not considered a classic endocrine disruptor, because it impacts reproductive and metabolic pathways primarily through interaction with the retinoid X receptor (RXR) and peroxisome proliferator-activated receptor (PPARγ) nuclear receptors.

Using assays recommended in the OECD CFEDTA, TBT was shown to alter the sex-ratio and to induce sperm abnormality in the Fish Sexual Development Test (FSDT) with zebrafish (McAllister and Kime 2003), and to delay frog development in the AMA (Shi et al. 2014), at low concentrations (0.1 and 10 ng/L, respectively). It also appeared positive in a FETAX-like assay (Guo et al. 2010). Based upon these findings, TBT would be identified as acting on endocrine pathways, although the specific MeOA (RXR and/or PPAR) would remain unknown. However, refinements to ToxCast™ now allow the identification of endocrine molecular initiating events through RXR and/or PPAR pathways. TBT activity for in vitro RXR and PPAR assays typically occurs at levels far less than those for baseline toxicity.

Interestingly, a more thorough evaluation of the available data clearly shows that TBT is highly toxic to a variety of aquatic taxa. Through a comparative analysis of the potency of TBT in various aquatic species, our review highlights the observation that fish are as sensitive, or more so, compared to molluscs when based on water exposure. This is an important conclusion because molluscs were long recognized as uniquely sensitive to this compound. TBT's precise MeOA is still incompletely understood but may include link/cross-talk between PPARs (i.e., carbohydrate, lipid, protein metabolism), RXRs (i.e., development), thyroid (growth) and even sex determination and differentiation pathways; the latter pathways may be stronger affected by TBT exposure in species where environmental factors play a significant role in determining sex ratios (e.g., zebrafish).

Current screening and assessment methodologies are able to identify TBT as a potent endocrine disruptor with a high environmental risk. If those approaches were available when TBT was introduced to the market, it is likely that its use would have been regulated sooner, thus avoiding the detrimental effects on marine gastropod populations and communities as documented over several decades.

This retrospective evaluation of TBT, a very potent endocrine disruptor in vertebrates and invertebrates, should serve as an example demonstrating how shortfalls within the framework of chemical toxicity evaluation can result in under-protective regulatory assessment. Nowadays, the assays included in the OECD Conceptual Framework, including those recently developed on gastropod molluscs would likely recognize TBT as a chemical of concern with respect to endocrine disruption, although its mechanism of action and potency across taxonomic groups would remain largely unknown. Reflective analysis of well-studied, but potentially misunderstood contaminants, such as TBT, provides important lessons that should serve as a guiding principle for future studies and refinements of assessment protocols.

11 Summary

Tributyltin (TBT) has been recognized as an endocrine disrupting chemical (EDC) for several decades. However, only in the last decade, was its primary endocrine mechanism of action (MeOA) elucidated—interactions with the nuclear retinoid-X receptor (RXR), peroxisome proliferator-activated receptor γ (PPARγ), and their heterodimers. This molecular initiating event (MIE) alters a range of reproductive, developmental, and metabolic pathways at the organism level. One of the most important lessons learned from years of research on TBT concerns apparent species sensitivity. Several aspects such as the rates of uptake and elimination, chemical potency, and metabolic capacity are all important for identifying the most sensitive species for a given chemical, including EDCs. As recognized for many years, TBT-induced responses are known to occur at very low concentrations for molluscs, a fact that has more recently also been observed in fish species. This review explores the MeOA and effects of TBT in different species (aquatic molluscs and other invertebrates, fish, amphibians, birds and mammals) according to the OECD Conceptual Framework for Endocrine Disruptor Testing and Assessment (CFEDTA). The information gathered on biological effects that are relevant for populations of aquatic animals was used to construct Species Sensitivity Distributions (SSDs) based on No Observed Effect Concentrations (NOECs) and Lowest Observed Effect Concentrations (LOECs). Fish appear at the lower end of these distributions, showing that they are as sensitive as molluscs, and for some species, even more sensitive. Concentrations in the range of 1 ng/L for water exposure (10 ng/g for whole-body burden) have been shown to elicit endocrine-type responses, whereas mortality occurs at water concentrations ten times higher. Current screening and assessment methodologies as compiled in the OECD CFEDTA are able to identify TBT as a potent endocrine disruptor with a high environmental risk for the original use pattern. If those approaches had been available when TBT was introduced to the market, it is likely that its use would have been regulated sooner, thus avoiding the detrimental effects on marine gastropod populations and communities as documented over several decades.

Acknowledgements The authors are indebted to the organizers of the SETAC Pellston Workshop® "Environmental Hazard and Risk Assessment Approaches for Endocrine-Active Substances" held in Pensacola, Florida in February 2016, for their support to the TBT case-study working group from which this review is derived.

Disclaimer

Any use of trade, firm, or product names is for descriptive purposes only and does not imply endorsement by the U.S. Government.

Conflict of Interest

The authors declare that they have no conflict of interest.

Annexes

Table 1 Updated OECD conceptual framework for testing and assessment of endocrine disrupters (OECD 2012a, b)

Level	Mammalian and non-mammalian toxicology
1	• Physical & chemical properties, e.g., MW reactivity, volatility, biodegradability
Existing data and non-test information	• All available (eco)toxicological data from standardized or non-standardized tests
	• Read across, chemical categories, QSARs and other in silico predictions, and ADME model predictions
2	• Estrogen or androgen receptor binding affinity

(continued)

Level	Mammalian and non-mammalian toxicology	
In vitro assays providing data about selected endocrine mechanism(s)/pathways(s)	• Estrogen receptor transactivation (TG 455–TG 457)	
	• Androgen or thyroid transactivation (if/when TGs are available)	
	• Steroidogenesis in vitro (TG 456)	
	• MCF-7 cell proliferation assays (ER ant/agonist)	
	• Other assays as appropriate	
	Mammalian toxicology	Non-mammalian toxicology
3	• Uterotrophic assay (TG 440)	• Xenopus embryo thyroid signaling assay (when/if TG is available)
In vivo assays providing data about selected endocrine mechanism(s)/pathway(s)[a]		• Amphibian metamorphosis assay (TG 231)
	• Hershberger assay (TG 441)	• Fish reproductive screening assay (TG 229)
		• Fish screening assay (TG 230)
		• Androgenized female stickleback screen (GD 140)
4	• Repeated dose 28-day study (TG 407)	• Fish sexual development test (TG 234)
In vivo assays providing data on adverse effects on endocrine relevant endpoints[b]	• Repeated dose 90-day study (TG 408)	• Fish reproduction partial life-cycle test (when/if TG is available)
	• One-generation reproduction toxicity study (TG 415)	• Larval amphibian growth & development assay (when TG is available)
	• Male pubertal assay (see GD 150, C4.3)[c]	
	• Female pubertal assay (see GD 150, C4.4)[c]	• Avian reproduction assay (TG 206)
		• Mollusc reproduction test (TG 242–TG 243, adopted 2016)[d]
	• Intact adult male endocrine screening assay (see GD 150, Annex 2.5)	• Chironomid toxicity test (TG 218–219)[d]
		• Daphnia reproduction test (with male induction) (TG 211)[d]
	• Prenatal developmental toxicity study (TG 414)	• Earthworm reproduction test (TG 222)[d]
	• Chronic toxicity and carcinogenicity studies (TG 451-3)	• Enchytraeid reproduction test (TG 220)[d]
		• Sediment water Lumbriculus toxicity test using spiked sediment (TG 225)[d]
	• Reproductive screening test (TG 421 if enhanced)	• Predatory mite reproduction test in soil (TG 226)[d]
	• Combined 28-day/reproductive screening assay (TG 422 if enhanced)	• Collembolan reproduction test in soil (TG 232)[d]
	• Developmental neurotoxicity (TG 426)	

(continued)

Level	Mammalian and non-mammalian toxicology	
5	• Extended one-generation reproductive toxicity study (TG 443)[e]	• Medaka extended one-generation reproduction test (MEOGRT) (TG 240)
In vivo assays providing more comprehensive data on adverse effects on endocrine relevant endpoints over more extensive parts of the life cycle of the organism[b]	• Two-generation reproduction toxicity study (TG 416 most recent update)	• FLCTT (fish life-cycle toxicity test) (when TG is available)
		• Avian two-generation reproductive toxicity assay (when TG is available)
		• Mysid life-cycle toxicity test (when TG is available)[d]
		• Copepod reproduction and development test (when TG is available)[d]
		• Sediment water chironomid life-cycle toxicity test (OECD TG 233)[d]
		• Mollusc full life-cycle assays (when TG is available)[d]

[a]Some assays may also provide some evidence of adverse effects
[b]Effects can be sensitive to more than one mechanism and may be due to non-ED mechanisms
[c]Depending on the guideline/protocol used, the fact that a substance may interact with a hormone system in these assays does not necessarily mean that when the substance is used it will cause adverse effects in humans or ecological systems
[d]At present, the available invertebrate assays solely involve apical endpoints which are able to respond to some endocrine disrupters and some non-EDs. Those in Level 4 are partial life-cycle tests, while those in Level 5 are full- or multiple life-cycle tests
[e]The Extended one-generation reproductive toxicity study (TG 443) is preferable for detecting endocrine disruption because it provides an evaluation of a number of endocrine endpoints in F1 juvenile and adult, which are not included in the two-generation study (TG 416 adopted 2001)

Table 2 LOEC (lowest observed effect concentration) of TBT for population-relevant endpoints in aquatic organisms

Species	Test conditions	Biological endpoint	LOEC (ng TBT/L)	Reported as[a]	Reference
Amphibians					
Xenopus laevis	AMA (stage 51 tadpoles)/CAMA (stage 46 tadpoles)	Embryo body length and development	10.00	nom	Shi et al. (2014)
Xenopus tropicalis	FETAX-like test	Tadpole development	50.00	nom	Guo et al. (2010)
Fish					
Carassius auratus	Adults, 54 days (semi-static exposure)	Body weight, swimming activity	2.15	mm	Zhang et al. (2016)
Danio rerio	Full life-cycle test	Sperm abnormality 3–5 months post-exposure	0.10	nom	McAllister and Kime (2003)
Danio rerio	Larvae, 0–70 days post-hatch	Sex ratio	0.10	nom	McAllister and Kime (2003)
Fundulus heteroclitus	Full life-cycle test	Sex ratio F_1	420.00	mm	Mochida et al. (2010)
Fundulus heteroclitus	Full life-cycle test	Time to hatch F_1	750.00	mm	Mochida et al. (2010)
Oryzias latipes	Embryos	Hatching success	10,440.00	mm?	Bentivegna and Piatkowski (1998)
Pimephales promelas	Early life-stage test, 32 days	Fry growth and weight	450.00	mm?	Brooke et al. (2003)
Poecilia reticulata	Male adults, 28 days	Reproductive behavior	5.00	nom	Tian et al. (2015)
Sebastiscus marmoratus	Embryos (gastrula stage), 144 h	Hatchability	10.00	nom	Zhang et al. (2011)
Crustaceans					
Acanthomysis sculpta	Life-cycle test	Reproduction	190.00	mm	Davidson et al. (2003)
Caprella danilevskii	Full life-cycle test	Sex ratio	100.00	nom	Ohji et al. (2002)
Daphnia magna	F_0 third-instar juveniles (3 days) and F_1 egg provisioning stage	Reproduction	88.00	mm	Jordão et al. (2015)
Daphnia magna	Adults, 21 days	Offspring/female	2500.00	nom	Oberdorster et al. (1998)

(continued)

Table 2 (continued)

Species	Test conditions	Biological endpoint	LOEC (ng TBT/L)	Reported as[a]	Reference
Pseudodiaptomus marinus	Full life-cycle test	F_0 fecundity (nauplii/female)	60.00	nom	Huang et al. (2006)
Pseudodiaptomus marinus	Full life-cycle test	Sex ratio F_1	20.00	nom	Huang et al. (2006)
Pseudodiaptomus marinus	Full life-cycle test	Ovigerous F_0 females (%)	60.00	nom	Huang et al. (2006)
Schmackeria poplesia	Full life-cycle test	Ovigerous females (%)	10.00	nom	Huang et al. (2010)
Schmackeria poplesia	Full life-cycle test	Larval development	60.00	nom	Huang et al. (2010)
Tigriopus japonicus	Ovigerous females, 14 days	Nauplii production	50.00	nom	Ara et al. (2010)
Molluscs					
Biomphalaria glabrata	From hatchlings to adults, prolonged exposure	Egg laying	1.00	nom	Ritchie et al. (2005)
Crassostrea gigas	Oyster spats, 28 days	Growth (weight gain)	5.00	nom?	Nell and Chvojka (1992)
Lymnaea stagnalis	Adults, 21 days	Polyembryony	46.36	mm	Giusti et al. (2013b)
Lymnaea stagnalis	Adults, 56 days	Fecundity	372.82	gmm	Charles et al. (2016)
Lymnaea stagnalis	Adults, 21 days	Growth (shell size)	481.47	mm	Giusti et al. (2013b)
Lymnaea stagnalis	Eggs, 21 days	Mean hatching time	100.00	nom	Bandow and Weltje (2012)
Lymnaea stagnalis	Adults, 170 days	Fecundity	1000.00	nom	Leung et al. (2007)
Nucella lapillus	Egg capsule to adults, 1 year	Reproduction	27.80	mm	Harding et al. (2003)
Marisa cornuarietis	Adults, 8 weeks	Embryo production	39.50	nom	Schulte-Oehlmann (1997)
Mercenaria mercenaria	Veliger larvae, 14 days static renewal	Growth (shell size)	5.00	nom	Laughlin et al. (1988)
Mytilus edulis	Larvae, 33 days	Growth (shell size)	50.00	nom	Lapota et al. (1993)

Potamopyrgus antipodarum	Adults, 28 days	Fecundity	184.71	mm	Ruppert et al. (2016)
Pinctada fucata martensii	One-week static renewal exposure of adult females	Embryo development	191.00	mm	Inoue et al. (2004)
Pinctada fucata martensii	24-h static exposure of fertilized eggs	Embryo development	192.00	mm	Inoue et al. (2004)
Ruditapes philippinarum	24-h static exposure of fertilized eggs	Embryo development	62.00	mm	Inoue et al. (2006)
Saccostrea commercialis	Oyster spats, 28 days	Growth (weight gain)	5.00	nom?	Nell and Chvojka (1992)
Scrobicularia plana	30-day static renewal exposure of 10-day old pediveliger larvae	Larval shell growth	30.00	mm	Ruiz et al. (1995)
Echinoderms					
Paracentrotus lividus	fertilized eggs, 48 h	Embryo development and larval length	200.00	nom	Bellas et al. (2005))
Ascidians					
Ciona intestinalis	Embryos (2-cell stage), 20 h	Embryo development	4000.00	nom	Bellas et al. (2005)
Ciona intestinalis	Exposure of oocytes and spermatozoa	Fertilization rate	290,000.00	nom	Gallo and Tosti (2013)

[a] *nom* nominal, *mm* arithmetic mean measured, *gmm* geometric mean measured

For each species, the lowest LOEC value was used to construct the SSD (Fig. 4)

Table 3 NOEC (no observed effect concentration) of TBT for population-relevant endpoints in aquatic organisms

Species	Test duration	Biological endpoint	NOEC (ng TBT/L)	Reported as[a]	Reference
Fish					
Cyprinodon variegatus	Juveniles, 24 days	Growth	340.00	nom?	De Bruijn et al. (2005)
Danio rerio	Full life-cycle test	Sperm abnormality 3–5 months post-exposure	0.01	nom	McAllister and Kime (2003)
Danio rerio	Larvae, 0–70 days post-hatch	Sex ratio	0.01	nom	McAllister and Kime (2003)
Fundulus heteroclitus	Full life-cycle test	Sex ratio F_1	70.00	mm	Mochida et al. (2010)
Fundulus heteroclitus	Full life-cycle test	Hatchability F_0 (%)	750.00	mm	Mochida et al. (2010)
Fundulus heteroclitus	Full life-cycle test	Time to hatch F_1	520.00	mm	Mochida et al. (2010)
Gasterosteus aculeatus	225 days	Reproduction	100.00	mm?	De Bruijn et al. (2005)
Oncorhynchus mykiss	Juveniles, 16 weeks	Growth	60.00	nom?	De Bruijn et al. (2005)
Pimephales promelas	Early life-stage test, 32 days	Fry growth and weight	150.00	mm?	Brooke et al. (2003)
Poecilia reticulata	Juveniles, 91 days	Growth	320.00	nom?	De Bruijn et al. (2005)
Sebastiscus marmoratus	Embryos (gastrula stage), 144 h	Hatchability	1.00	nom	Zhang et al. (2011)
Crustaceans					
Acanthomysis sculpta	Life-cycle test	Reproduction	90.00	mm	Davidson et al. (2003)
Acartia tonsa	Eggs, 8 days	Larval development EC_{10}	0.70	nom	Kusk and Petersen (1997)
Caprella danilevskii	Full life-cycle test	Sex ratio	10.00	nom	Ohji et al. (2002)
Daphnia magna	Adults, 21 days	Reproduction	160.00	mm?	Kühn et al. (2005)
Daphnia magna	Adults, 21 days	Offspring/female	1250.00	nom	Oberdorster et al. (1998)

Eurytemora affinis	Adults, 13 days	Reproduction	10.00	mm?	De Bruijn et al. (2005)
Eurytemora affinis	Egg-carrying females, 13 days	Brood size	224.00	mm	Hall et al. (2003)
Pseudodiaptomus marinus	Full life-cycle test	Sex ratio F_1	6.00	nom	Huang et al. (2006)
Pseudodiaptomus marinus	Full life-cycle test	Ovigerous F_0 females (%)	20.00	nom	Huang et al. (2006)
Pseudodiaptomus marinus	Full life-cycle test	F_0 fecundity (nauplii/female)	20.00	nom	Huang et al. (2006)
Schmackeria poplesia	Full life-cycle test	Ovigerous females (%)	5.00	nom	Huang et al. (2010)
Schmackeria poplesia	Full life-cycle test	Larval development	40.00	nom	Huang et al. (2010)
Tigriopus japonicus	Ovigerous females, 14 days	Nauplii production	25.00	nom	Ara et al. (2010)
Molluscs					
Isognomon californicum	Gametes 48 h	Fertilization rate	1000.00	nom	Ringwood (1992)
Isognomon californicum	Embryos, 48 h	Embryo development	100.00	nom	Ringwood (1992)
Isognomon californicum	Veliger larvae, 4 days	Larval growth	20.00	nom	Ringwood (1992)
Lymnaea stagnalis	Adults, 170 days	Fecundity	10.00	nom	Leung et al. (2007)
Lymnaea stagnalis	Adults, 170 days	Population growth rate	2745.00	nom	Leung et al. (2007)
Lymnaea stagnalis	Adults, 56 days	Fecundity	231.00	gmm	Charles et al. (2016)
Lymnaea stagnalis	Adults, 21 days	Growth (shell size)	229.74	mm	Giusti et al. (2013a)
Lymnaea stagnalis	Eggs, 21 days	Mean hatching time	30.00	nom	Bandow and Weltje (2012)
Nucella lapillus	Egg capsule to adults, 1 year	Reproduction	7.40	mm	Harding et al. (2003)
Mytilus edulis	Larvae, 33 days	Growth (shell size)	6.00	nom	Lapota et al. (1993)
Pinctada fucata martensii	One-week static renewal exposure of adult females	Embryo development	92.00	mm	Inoue et al. (2004)

(continued)

Table 3 (continued)

Species	Test duration	Biological endpoint	NOEC (ng TBT/L)	Reported as[a]	Reference
Pinctada fucata martensii	24-h static exposure of fertilized eggs	Embryo development	91.00	mm	Inoue et al. (2004)
Potamopyrgus antipodarum	Adults, 28 days	Fecundity	95.65	mm	Ruppert et al. (2016)
Echinoderms					
Echinometra mathaei	Gametes, 60–90 min	Fertilization rate	1000.00	nom	Ringwood (1992)
Ophioderma brevispina	28 days	Regeneration	10.00	mm?	Walsh et al. (2005)
Paracentrotus lividus	Fertilized eggs, 48 h	Embryo development and larval length	100.00	nom	Bellas et al. (2005)
Ascidians					
Ciona intestinalis	Embryos (2-cell stage), 20 h	Embryo development	2000.00	nom	Bellas et al. (2005)
Annelids					
Neanthes arenaceodentata	Adults, 70 days	Growth	50.00	nom?	Moore et al. (2003)

[a]*nom* nominal, *mm* arithmetic mean measured, *gmm* geometric mean measured

For each species, the lowest NOEC value was used to construct the SSD (Fig. 5)

Table 4 LOEC (lowest observed effect concentration) of TBT for mortality in aquatic organisms

Species	Test duration	Biological endpoint	LOEC (ng TBT/L)	Reported as[a]	Reference
Amphibians					
Xenopus tropicalis	FETAX-like test, 48 h	Tadpole development	50.00	nom	Guo et al. (2010)
Fish					
Cyprinodon variegatus	n.r.	Parental survival	560.00	mm?	United States Environmental Protection Agency - US EPA (2008)
Oryzias latipes	Embryonic stages, 96 h	Embryo survival	41,500.00	mm?	Bentivegna and Piatkowski (1998)
Crustaceans					
Acartia tonsa	Nauplii larvae, 6 days	Larval survival	23.50	mm	Bushong et al. (1990)
Caprella danilevskii	Full life-cycle test	Embryo survival	10.00	nom	Ohji et al. (2003b)
Daphnia magna	21-day chronic	Survival and reproduction	200.00	mm?	Brooke et al. (2003))
Daphnia magna	21-day chronic	Survival and reproduction	340.00	mm?	ABC Laboratories Inc. (2003)
Daphnia magna	21-day chronic run over two generations	Adult survival	2225.00	nom	Oberdorster et al. (1998)
Eurytemora affinis	n.r.	Neonate survival	88.00	mm?	Hall et al. (2003)
Palaemon serratus	Zoe I stage larvae, 48 h	Larval survival	62,500.00	nom	Bellas et al. (2005)
Pseudodiaptomus marinus	Nauplii to copepodites, 13 days	Larvae to adult survival F_0	60.00	nom	Huang et al. (2006)
Schmackeria poplesia	Full life-cycle test	Adult survival	20.00	nom	Huang et al. (2010)
Tisbe biminiensis	Adult (7–10 day-old), 48 h	Adult survival	34,000.00	nom	Varella Motta da Costa et al. (2014)
Molluscs					
Lymnaea stagnalis	Adults, 170 days	Adult survival	1000.00	nom	Leung et al. (2007)
Lymnaea stagnalis	Adults, 170 days	Juvenile survival	1000.00	nom	Leung et al. (2007)

(continued)

Table 4 (continued)

Species	Test duration	Biological endpoint	LOEC (ng TBT/L)	Reported as[a]	Reference
Mytilus edulis	33 days	Larval survival	50.00	nom	Lapota et al. (1993)
Ruditapes philippinarum	Static renewal exposure of veliger larvae (D-larvae stage) for 13 days	Survival and development of veliger larvae	130.00	mm	Inoue et al. (2007)
Scrobicularia plana	30-day static renewal exposure of 10-day-old pediveliger larvae	Larval survival	56.10	mm	Ruiz et al. (1995)
Insects					
Chironomus riparius	Fourth instar larvae, 48 h	Larval survival	48.80	nom	Hahn and Schulz (2002)
Annelids					
Hydroides elegans	Early development (egg to juvenile)	Adult (post-spawning female) survival	10,000.00	nom	Lau et al. (2007)
Hydroides elegans	Early development (egg to juvenile)	Juvenile survival	1000.00	nom	Lau et al. (2007)
Neanthes arenaceodentata	Adults, 70 days	Adult survival	500.00	nom?	Moore et al. (2003)

[a]*nom* nominal, *mm* arithmetic mean measured, *gmm* geometric mean measured

For each species, the lowest LOEC value was used to construct the SSD (Fig. 7). Species for which LOEC values were higher than 1000 ng/L were not considered sensitive and were not included in the SSD

References

ABC Laboratories Inc. (2003) 1990 in United States Environmental Protection Agency. Ambient aquatic life water quality criteria for tributyltin (TBT) – final. Office of Water, Office of Science and Technology Health and Ecological Criteria Division, Washington, DC. 129 pages

Abidli S, Santos MM, Lahbib Y, Castro LFC, Reis-Henriques MA, El Menif NT (2012) Tributyltin (TBT) effects on *Hexaplex trunculus* and *Bolinus brandaris* (Gastropoda: Muricidae): imposex induction and sex hormone levels insights. Ecol Indic 13:13–21

Abidli S, Castro LFC, Lahbib Y, Reis-Henriques MA, El Meni NT, Santos MM (2013) Imposex development in *Hexaplex trunculus* (Gastropoda: Caenogastropoda) involves changes in the transcription levels of the retinoid X receptor (RXR). Chemosphere 93:1161–1167

Alzieu C (2000) Environmental impact of TBT: the French experience. Sci Total Environ 258:99–102

Alzieu C, Sanjuan J, Deltriel J-P, Borel M (1986) Tin contamination in Arcachon bay: effects on oyster shell anomalies. Mar Pollut Bull 17:494–498

Ankley GT, Bennett RS, Erickson RJ, Hoff DJ, Hornung MW, Johnson RD, Mount DR, Nichols JW, Russom CL, Schmieder PK, Serrano JA, Tietge JE, Villeneuve DL (2010) Adverse Outcome Pathways: a conceptual framework to support ecotoxicology research and risk assessment. Environ Toxicol Chem 29:730–741

Antizar-Ladislao B (2008) Environmental levels, toxicity and human exposure to tributyltin (TBT)-contaminated marine environment. Environ Int 34:292–308

Arnold CG, Weidenhaupt A, David MM, Müller SR, Haderlein SB, Schwarzenbach RP (1997) Aqueous speciation and 1-octanol-water partitioning of tributyl- and triphenyltin: effect of pH and ion composition. Environ Sci Technol 31:2596–2602

Avaca MS, Martin P, van der Molen S, Narvarte M (2015) Comparative study of the female gametogenic cycle in three populations of *Buccinanops globulosus* (Caenogastropoda: Nassariidae) from Patagonia. Helgol Mar Res 69:87–99

Bailey SK, Davies IM, Harding MJC (1995) Tributyltin contamination and its impact on *Nucella lapillus* populations. Proc Roy Soc Edinb B Biol Sci 103:113–126

Bandow C, Weltje L (2012) Development of an embryo toxicity test with the pond snail *Lymnaea stagnalis* using the model substance tributyltin and common solvents. Sci Total Environ 435–436:90–95

Barroso CM, Moreira MH, Bebianno MJ (2002) Imposex, female sterility and organotin contamination of the prosobranch *Nassarius reticulatus* from the Portuguese coast. Mar Ecol Prog Ser 230:127–135

Bartlett AJ, Borgmann U, Dixon DG, Batchelor SP, Maguire RJ (2007) Comparison of toxicity and bioaccumulation of tributyltin in Hyalella azteca and five other freshwater invertebrates. Water Qual Res J Can 42:1–10

Batley GE, Scammell MS, Brockbank CI (1992) The impact of the banning of tributyltin-based antifouling paints on the Sydney rock oyster, *Saccostrea commercialis*. Sci Total Environ 122:301–314

Belcher SM, Cookman CJ, Patisaul HB, Stapleton HM (2014) *In vitro* assessment of human nuclear hormone receptor activity and cytotoxicity of the flame retardant mixture FM 550 and its triarylphosphate and brominated components. Toxicol Lett 228:93–102

Bentivegna CS, Piatkowski T (1998) Effects of tributyltin on medaka (*Oryzias latipes*) embryos at different stages of development. Aquat Toxicol 44:117–128

Birch GF, Scammell MS, Besley CH (2014) The recovery of oyster (*Saccostrea glomerata*) populations in Sydney estuary (Australia). Environ Sci Pollut Res Int 21:766–773

Birchenough AC, Evans SM, Moss C, Welch R (2002) Re-colonisation and recovery of populations of dogwhelks *Nucella lapillus* (L.) on shores formerly subject to severe TBT contamination. Mar Pollut Bull 44:652–659

Blackmore G (2000) Imposex in *Thais clavigera* (Neogastropoda) as an indicator of TBT (tributyltin) bioavailability in coastal waters of Hong Kong. J Moll Stud 66:1–8

Blomhoff R, Blomhoff HK (2006) Overview of retinoid metabolism and function. J Neurobiol 66:606–630

Boulahtouf A, Grimaldi M, Coutellec M-A, Besnard A-L, Echasseriau Y, Bourguet W, Balaguer P, Lagadic L (2015) Ligand affinity of the *Lymnaea stagnalis* estrogen and retinoid-X receptors (LsER and LsRXR): implications for detecting endocrine disruptors. 25th SETAC Europe annual meeting, Barcelona, extended abstract

Bouton D, Escriva H, de Mendonca RL, Glineur BB, Noel C, Robinson-Rechavi M, de Groot A, Cornette J, Laudet V, Pierce RJ (2005) A conserved retinoid X receptor (RXR) from the mollusk *Biomphalaria glabrata* transactivates transcription in the presence of retinoids. J Mol Endocrinol 34:567–582

Bray S, McVean EM, Nelson A, Herbert RJH, Hawkins SJ, Hudson MD (2012) The regional recovery of *Nucella lapillus* populations from marine pollution, facilitated by man-made structures. J Mar Biol Assoc UK 92:1585–1594

Bryan GW, Gibbs PE, Burt GR, Hummerstone LG (1987) The effects of tributyltin (TBT) accumulation on adult dogwhelks, *Nucella lapillus*: long term field and laboratory experiments. J Mar Biol Assoc UK 67:525–544

Bryan GW, Burt GR, Gibbs PE, Pascoe PL (1993) *Nassarius reticulatus* (Nassariidae: Gastropoda) as an indicator of tributyltin pollution before and after TBT restrictions. J Mar Biol Assoc UK 73:913–929

Campo-Paysaa F, Marlétaz F, Laudet V, Schubert M (2008) Retinoic acid signaling in development: tissue-specific functions and evolutionary origins. Genesis 46:640–656

Castro LFC, Lima D, Machado A, Melo C, Hiromori Y, Nishikawa J, Nakanishi T, Reis-Henriques MA, Santos MM (2007) Imposex induction is mediated through the retinoid X receptor signaling pathway in the neogastropod *Nucella lapillus*. Aquat Toxicol 85:57–66

Chamorro-García R, Sahu M, Abbey RJ, Laude J, Pham N, Blumberg B (2013) Transgenerational inheritance of increased fat depot size, stem cell reprogramming, and hepatic steatosis elicited by prenatal exposure to the obesogen tributyltin in mice. Environ Health Perspect 121:359–366

Champ MA (2000) A review of organotin regulatory strategies, pending actions, related costs and benefits. Sci Total Environ 258:21–71

Chang X, Kobayashi T, Senthilkumaran B, Kobayashi-Kajura H, Sudhakumari CC, Nagahama Y (2005) Two types of aromatase with different encoding genes, tissue distribution and developmental expression in Nile tilapia (Oreochromis niloticus) Gen Comp Endocrinol 141:101–115

Cheshenko K, Pakdel F, Segner H, Kah O, Eggen RI (2008) Interference of endocrine disrupting chemicals with aromatase CYP19 expression or activity, and consequences for reproduction of teleost fish. Gen Comp Endocrinol 155:31–62

Choi MJ, Kim SC, Kim AN, Kwon HB, Ahn RS (2007) Effect of endocrine disruptors on the oocyte maturation and ovulation in amphibians, *Rana dybowskii*. Integr Biosci 11:1–8

Choudhury AKR (2014) Environmental impacts of the textile industry and its assessment through life cycle management. In: Muthu SS (ed) Roadmap to sustainable textiles and clothing. Environmental and social aspects of textiles and clothing supply chain. Springer, Singapore, pp 1–40

Coenen TMM, Brouwer A, Enninga IC, Koeman JH (1992) Subchronic toxicity and reproduction effects of tri-n-butyltin oxide in Japanese quail. Arch Environ Contam Toxicol 23:457–463

Coenen TMM, Enninga IC, Cave DA, Hoeven JCM (1994) Hematology and serum biochemistry of Japanese quail fed dietary tri-n-butyltin oxide during reproduction. Arch Environ Contam Toxicol 26:227–233

Colliar L, Sturm A, Leaver MJ (2011) Tributyltin is a potent inhibitor of piscine peroxisome proliferator-activated receptor alpha and beta. Comp Biochem Physiol C 153:168–173

Cooke GM (2002) Effect of organotins on human aromatase activity *in vitro*. Toxicol Lett 126:121–130

Cooke GM, Forsyth DS, Bondy GS, Tachon R, Tague B, Coady L (2008) Organotin speciation and tissue distribution in rat dams, fetuses, and neonates following oral administration of tributyltin chloride. J Toxicol Environ Health A 71:384–395

Couceiro L, Diaz J, Albaina N, Barreiro R, Irabien JA, Ruiz JM (2009) Imposex and gender-independent butyltin accumulation in the gastropod *Nassarius reticulatus* from the Cantabrian coast (N Atlantic Spain). Chemosphere 76:424–427

Decherf S, Seugnet I, Fini J-B, Clerget-Froidevaux M-S, Demeneix BA (2010) Disruption of thyroid hormone-dependent hypothalamic set-points by environmental contaminants. Mol Cell Endocrinol 323:172–182

Duft M, Schmitt C, Bachmann J, Brandelik C, Schulte-Oehlmann U, Oehlmann J (2007) Prosobranch snails as test organisms for the assessment of endocrine active chemicals—an overview and a guideline proposal for a reproduction test with the freshwater mudsnail *Potamopyrgus antipodarum*. Ecotoxicology 16:169–182

ECHA (2008) Member state committee support document for identification of Bis(tributyltin) oxide as a substance of very high concern. European Chemicals Agency, Oct 2008

Elliott JE, Harris ML, Wilson LK, Smith BD, Batchelor SP, Maguire J (2007) Butyltins, trace metals and morphological variables in surf scoter (*Melanitta perspicillata*) wintering on the south coast of British Columbia, Canada. Environ Pollut 149:114–124

European Union (EU) (2005) Common implementation strategy for the water framework directive: environmental quality standards (EQS), Substance Data Sheet Priority Substance No 30: Tributyltin compounds (TBT-ion) Final version, Brussels, Jan 2005

Faqi SA, Hilbig V, Pfeil R, Solecki R (1999) Dietary TBTO exposure to the Japanese quail: relation between exposure period and appearance of reproductive effects. Bull Environ Contam Toxicol 63:415–422

Fent K (1996) Ecotoxicology of organotin compounds. Crit Rev Toxicol 26:1–117

Fent K, Hunn J (1996) Cytotoxicity of organic environmental chemicals to fish liver cells (PLHC-1). Mar Environ Res 42:377–382

Gagné F, Blaise C, Salazar M, Salazar S, Hansen PD (2001) Evaluation of estrogenic effects of municipal effluents to the freshwater mussel *Elliptio complanata*. Comp Biochem Physiol C 128:213–225

Gibbs PE (1996) Oviduct as a sterilising effect of tributyltin (TBT)-induced imposex in *Ocenebra erinacea* (Gastropoda: Muricidae). J Moll Stud 62:403–413

Gibbs PE (2009) Long-term tributyltin (TBT)-induced sterilization of neogastropods: persistence of effects in Ocenebra erinacea over 20 years in the vicinity of Falmouth (Cornwall, UK). J Mar Biol Assoc UK 89:135–138

Gibbs PE, Bryan GW (1986) Reproductive failure in populations of the dogwhelk, *Nucella lapillus*, caused by imposex induced by tributyltin from antifouling paints. J Mar Biol Assoc UK 66:767–777

Gibbs PE, Bryan GW (1996a) Reproductive failure in the gastropod *Nucella lapillus* associated with imposex caused by tributyltin pollution: a review. In: Champ MA, Seligman PF (eds) Organotin. Chapman and Hall, London, UK, pp 259–280

Gibbs PE, Bryan GW (1996b) TBT-induced imposex in neogastropod snails: masculinization to mass extinction. In: De Mora SJ (ed) Tributyltin: case study of an environmental contaminant. Cambridge University Press, New York, NY, USA, pp 212–236

Gibbs PE, Bryan GW, Pascoe PL, Burt GR (1987) The use of the dog-whelk, *Nucella lapillus,* as an indicator of tributyltin (TBT) contamination. J Mar Biol Assoc UK 67:507–523

Gibbs PE, Pascoe PL, Burt GR (1988) Sex change in the female dog-whelk, *Nucella lapillus*, induced by tributyltin from antifouling paints. J Mar Biol Assoc UK 68:715–731

Giusti A (2013) Impacts and mechanisms of action of endocrine disrupting chemicals on the hermaphroditic freshwater gastropod *Lymnaea stagnalis* (Linnaeus, 1758). PhD thesis, University of Liège, Belgium, 244 p

Giusti A, Joaquim-Justo C (2013) Esterification of vertebrate-like steroids in molluscs: a target of endocrine disruptors? Comp Biochem Physiol C 158:187–198

Giusti A, Barsi A, Dugué M, Collinet M, Thome JP, Joaquim-Justo C, Roig B, Lagadic L, Ducrot V (2013a) Reproductive impacts of tributyltin (TBT) and triphenyltin (TPT) in the hermaphroditic freshwater gastropod *Lymnaea stagnalis*. Environ Toxicol Chem 32:1552–1560

Giusti A, Ducrot V, Joaquim-Justo C, Lagadic L (2013b) Testosterone levels and fecundity in the hermaphroditic aquatic snail *Lymnaea stagnalis* exposed to testosterone and endocrine disruptors. Environ Toxicol Chem 32:1740–1745

Gooding MP, LeBlanc GA (2001) Biotransformation and disposition of testosterone in the eastern mud snail *Ilyanassa obsoleta*. Gen Comp Endocrinol 122:172–180

Gooding MP, Wilson VS, Folmar LC, Marcovich DT, LeBlanc GA (2003) The biocide tributyltin reduces the accumulation of testosterone as fatty acid esters in the mud snail (*Ilyanassa obsoleta*). Environ Health Perspect 111:426–430

Grimaldi M, Boulahtouf A, Delfosse V, Thouennon E, Bourguet W, Balaguer P (2015) Reporter cell lines for the characterization of the interactions between human nuclear receptors and endocrine disruptors. Front Endocrinol 6:62. https://doi.org/10.3389/fendo.2015.00062

Grün F, Watanabe H, Zamanian Z, Maeda L, Arima K, Cubacha R, Gardiner DM, Kanno J, Iguchi T, Blumberg B (2006) Endocrine-disrupting organotin compounds are potent inducers of adipogenesis in vertebrates. Mol Endocrinol 20:2141–2155

Guo S, Quan L, Shi H, Barry T, Cao Q, Liu J (2010) Effects of tributyltin (TBT) on *Xenopus tropicalis* embryos at environmentally relevant concentrations. Chemosphere 79:529–533

Gutierrez-Mazariegos J, Nadendla EK, Lima D, Pierzchalski K, Jones JW, Kane M, Nishikawa JI, Hiromori Y, Nakanishi T, Santos MM, Castro LFC, Bourguet W, Schubert M, Laudet V (2014) A mollusk retinoic acid receptor (RAR) ortholog sheds light on the evolution of ligand binding. Endocrinology 155:4275–4286

Hahn T, Schulz R (2002) Ecdysteroid synthesis and imaginal disc development in the midge *Chironomus riparius* as biomarkers for endocrine effects of tributyltin. Environ Toxicol Chem 21:1052–1057

Hano T, Oshima Y, Kim SG, Satone H, Oba Y, Kitano T, Inoue S, Shimasaki Y, Honjo T (2007) Tributyltin causes abnormal development in embryos of medaka, *Oryzias latipes*. Chemosphere 69:927–933

Harazono A, Ema M, Ogawa Y (1996) Pre-implantation embryonic loss induced by tributyltin chloride in rats. Toxicol Lett 89:185–190

Harazono A, Ema M, Ogawa Y (1998) Evaluation of early embryonic loss induced by Tributyltin chloride in rats: phase- and dose-dependent antifertility effects. Arch Environ Contam Toxicol 34:94–99

Harding MJC, Rodger GK, Davies IM, Moore JJ (1997) Partial recovery of the dogwhelk (*Nucella lapillus*) in Sullom Voe, Shetland from tributyltin contaminations. Mar Environ Res 44:285–304

Haubruge E, Petit F, Gage MJ (2000) Reduced sperm counts in guppies (*Poecilia reticulata*) following exposure to low levels of tributyltin and bisphenol A. Proc Roy Soc Lond B 67:2333–2337

Hawkins SJ, Proud SV, Spence SK, Southward AJ (1994) From the individual to the community and beyond: water quality, stress indicators and key species in coastal waters. In: Sutcliffe DW (ed) Water quality & stress indicators in marine and freshwater systems: linking levels of organisation. Ambleside, UK, Freshwater Biological Association, pp 35–62

Heidrich DD, Steckelbroeck S, Klingmuller D (2001) Inhibition of human cytochrome P450 aromatase activity by butyltins. Steroids 66:763–769

Ho KKY, Zhou G-J, EGB X, Wang X, Leung KMY (2016) Long-term spatio-temporal trends of organotin contaminations in the marine environment of Hong Kong. PLoS One 11(5): e0155632

Holbech H, Kinnberg K, Petersen GI, Jackson P, Hylland K, Norrgren L, Bjerregaard P (2006) Detection of endocrine disrupters: evaluation of a fish sexual development test (FSDT). Comp Biochem Physiol C 144:57–66

Horiguchi T, Shiraishi H, Shimizu M, Morita M (1994) Imposex and organotin compounds in *Thais clavigera* and *T. bronni* in Japan. J Mar Biol Assoc UK 74:651–669

Huang Y, Zhu L, Liu G (2006) The effects of bis(tributyltin) oxide on the development, reproduction and sex ratio of calanoid copepod *Pseudodiaptomus marinus*. Estuar Coast Shelf Sci 69:147–152

Huang Y, Zhu LY, Qiu XC, Zhang TW (2010) Effect of bis(tributyltin) oxide on reproduction and population growth rate of calanoid copepod *Schmackeria poplesia*. Chinese J Oceanol Limnol 28:280–287

Iguchi T, Katsu Y (2008) Commonality in signaling of endocrine disruption from snail to human. Bioscience 58:1061–1067

Inoue S, Oshima Y, Nagai K, Yamamoto T, Go J, Kai N, Honjo T (2004) Effect of maternal exposure to tributyltin on reproduction of the pearl oyster (Pinctada fucata martensii). Environ Toxicol Chem 23:1276–1281

Inoue S, Oshima Y, Usuki H, Hamaguchi M, Hanamura Y, Kai N, Shimasaki Y, Honjo T (2006) Effects of tributyltin maternal and/or waterborne exposure on the embryonic development of the Manila clam, *Ruditapes philippinarum*. Chemosphere 63:881–888

Janer G, Lavado R, Thibaut R, Porte C (2004) Effects of 17b-estradiol exposure in the mussel *Mytilus galloprovincialis*. Mar Environ Res 58:443–446

Jordão R, Casas J, Fabrias G, Campos B, Piña B, Lemos MF, Soares AM, Tauler R, Barata C (2015) Obesogens beyond vertebrates: lipid perturbation by tributyltin in the crustacean *Daphnia magna*. Environ Health Perspect 123:813–819

Kanayama T, Kobayashi N, Mamiya S, Nakanishi T, Nishikawa J (2005) Organotin compounds promote adipocyte differentiation as agonists of the peroxisome proliferator-activated receptor γ/retinoid-X receptor pathway. Mol Pharmacol 67:766–774

Kanda A, Takahashi T, Satake H, Minakata H (2006) Molecular and functional characterization of a novel gonadotropin-releasing-hormone receptor isolated from the common octopus (*Octopus vulgaris*). Biochem J 395:125–135

Kaur S, Jobling S, Jones CS, Noble LR, Routledge EJ, Lockyer AE (2015) The nuclear receptors of Biomphalaria glabrata and Lottia gigantea: implications for developing new model organisms. PLoS One 10(4)

Kazeto Y, Ijiri S, Place AR, Zohar Y, Trant JM (2001) The 5′-flanking regions of CYP19A1 and CYP19A2 in zebrafish. Biochem Biophys Res Comm 288:503–508

Kim NS, Hong SH, Yim UH, Shin K-H, Shim WJ (2014) Temporal changes in TBT pollution in water, sediment, and oyster from Jinhae Bay after the total ban in South Korea. Mar Pollut Bull 86:547–554

Kinnberg K, Holbech H, Petersen GI, Bjerregaard P (2007) Effects of the fungicide prochloraz on the sexual development of zebrafish (*Danio rerio*). Comp Biochem Physiol C 145:165–170

Kirchner S, Kieu T, Chow C, Casey S, Blumberg B (2010) Prenatal exposure to the environmental obesogen tributyltin predisposes multipotent stem cells to become adipocytes. Mol Endocrinol 24:526–539

Kishida M, Nakao R, Novillo A, Callard IP, Osada M (2005) Molecular cloning and expression analysis of cDNA fragments related to estrogen receptor from blue mussel, *Mytilus edulis*. Proc Jpn Soc Comp Endocrinol 20:75

Kishta O, Adeeko A, Li D, Luu T, Brawer JR, Morales C, Hermo L, Robaire B, Hales BF, Barthelemy J, Cyr DG, Trasler JM (2007) In utero exposure to tributyltin chloride differentially alters male and female fetal gonad morphology and gene expression profiles in the Sprague–Dawley rat. Reprod Toxicol 23:1–11

Kortner TM, Pavlikova N, Arukwe A (2010) Effects of tributyltin on salmon interrenal CYP11 beta, steroidogenic factor-1 and glucocorticoid receptor transcripts in the presence and absence of second messenger activator, forskolin. Mar Environ Res 69:S56–S58

Kusk KO, Petersen S (1997) Acute and chronic toxicity of tributyltin and linear alkylbenzene sulfonate to the marine copepod *Acartia tonsa*. Environ Toxicol Chem 16:1629–1633

Laflamme L, Hamann G, Messier N, Maltais S, Langlois M-F (2002) RXR acts as a coregulator in the regulation of genes of the hypothalamo-pituitary axis by thyroid hormone receptors. J Mol Endocrinol 29:61–72

Lagerström M, Strand J, Eklund Y (2016) Total tin and organotin speciation in historic layers of antifouling paint on leisure boat hulls. Environ Pollut 220(B):1333–1341

LeBlanc GA, Gooding MP, Sternberg RM (2005) Testosterone-fatty acid esterification: a unique target for the endocrine toxicity of tributyltin to gastropods. Integr Comp Biol 45:81–87

Leung KMY, Kwong RPY, Ng WC, Horiguchi T, Qiu JW, Yang RQ, Song MY, Jiang GB, Zheng GJ, Lam PKS (2006) Ecological risk assessments of endocrine disrupting organotin compounds using marine neogastropods in Hong Kong. Chemosphere 65:922–938

Leung KMY, Grist EPM, Morley NJ, Morritt D, Crane M (2007) Chronic toxicity of tributyltin to development and reproduction of the European freshwater snail *Lymnaea stagnalis* (L.) Chemosphere 66:1358–1366

Lewis JA, Baran IJ, Carey JM, Fletcher LE (2010) A contaminant in decline: long-term TBT monitoring at a naval base in Western Australia. Aust J Ecotoxicol 16:17–34

Lilley TM, Ruokolainen L, Pikkarainen A, Laine VN, Kilpimaa J, Rantala MJ, Nikinmaa M (2012) Impact of tributyltin on immune response and life history traits of *Chironomus riparius*: single and multigeneration effects and recovery from pollution. Environ Sci Technol 46:7382–7389

Lima D, Reis-Henriques MA, Silva R, Santos AI, Castro LFC, Santos MM (2011) Tributyltin-induced imposex in marine gastropods involves tissue-specific modulation of the retinoid X receptor. Aquat Toxicol 101:221–227

Lima D, Castro LFC, Coelho I, Lacerda R, Gesto M, Soares J, André A, Capela R, Torres T, Carvalho AP, Santos MM (2015) Effects of tributyltin and other retinoid receptor agonists in reproductive-related endpoints in the zebrafish (*Danio rerio*). J Toxicol Environ Health A 78:747–760

Maack G, Segner H (2003) Morphological development of the gonads in zebrafish. J Fish Biol 62:895–906

le Maire A, Grimaldi M, Roecklin D, Dagnino S, Vivat-Hannah V, Balaguer P, Bourguet W (2009) Activation of RXR-PPAR heterodimers by organotin environmental endocrine disruptors. EMBO Rep 10:367–373

Matthiessen P, Gibbs PE (1998) Critical appraisal of the evidence for tributyltin-mediated endocrine disruption in molluscs. Environ Toxicol Chem 17:37–43

Matthiessen P, Weltje L (2015) A review of the effects of azole compounds in fish and their possible involvement in masculinization of wild fish populations. Crit Rev Toxicol 45:453–467

McAllister BG, Kime DE (2003) Early life exposure to environmental levels of the aromatase inhibitor tributyltin causes masculinisation and irreversible sperm damage in zebrafish (*Danio rerio*). Aquat Toxicol 65:309–316

McCarty LS, Landrum PF, Luoma SN, Meador JP, Merten AA, Shephard BK, van Wezel AP (2011) Advancing environmental toxicology through chemical dosimetry: external exposures versus tissue residues. Integr Environ Assess Manag 7:7–27

McElroy AE, Barron MG, Beckvar N, Kane Driscoll SB, Meador JP, Parkerton TF, Preuss TG, Steevens JA (2011) A review of the tissue residue approach for organic and organometallic compounds in aquatic organisms. Integr Environ Assess Manag 7:50–74

McGinnis CL, Crivello JF (2011) Elucidating the mechanism of action of tributyltin (TBT) in zebrafish. Aquat Toxicol 103:25–31

McGinnis CL, Encarnacao PC, Crivello JF (2012) Dibutyltin (DBT): an endocrine disrupter in zebrafish. J Exp Mar Biol Ecol 430–431:43–47

McVey MJ, Cooke GM (2003) Inhibition of rat testis microsomal 3β-hydroxysteroid dehydrogenase activity by tributyltin. J Steroid Biochem Mol Biol 86:99–105

Meador JP (1997) Comparative toxicokinetics of tributyltin in five marine species and its utility in predicting bioaccumulation and toxicity. Aquat Toxicol 37:307–326

Meador JP (2000) Predicting the fate and effects of tributyltin in marine systems. Rev Environ Contam Toxicol 166:1–48

Meador JP (2006) Rationale and procedures for using the tissue-residue approach for toxicity assessment and determination of tissue, water, and sediment quality guidelines for aquatic organisms. Hum Ecol Risk Assess 12:1018–1073

Meador JP (2011) Organotins in aquatic biota: occurrence in tissue and toxicological significance. In: Beyer WN, Meador JP (eds) Environmental contaminants in biota: interpreting tissue concentrations. Taylor and Francis, Boca Raton, FL, pp 255–284

Meador JP, McCarty LS, Escher BI, Adams WJ (2008) The tissue-residue approach for toxicity assessment: concepts, issues, application, and recommendations. J Environ Monit 10:1486–1498

Meador JP, Sommers FC, Cooper K, Yanagida G (2011) Tributyltin and the obesogen metabolic syndrome in a salmonid. Environ Res 111:50–56

Mengeling BJ, Murk AJ, Furlow JD (2016) Trialkyltin retinoid-X receptor agonists selectively potentiate thyroid hormone induced programs of Xenopus laevis metamorphosis. Endocrinology 157:2712–2723

Mizukawa H, Takahashi S, Nakayama K, Sudo A, Tanabe S (2009) Contamination and accumulation feature of organotin compounds in common cormorants (*Phalacrocorax carbo*) from Lake Biwa, Japan. In: Obayashi Y, Isobe T, Subramanian A, Suzuki S, Tanabe S (eds) Interdisciplinary studies on environmental chemistry — environmental research in Asia. TERRAPUB, pp 153–161

Mochida K, Ito K, Kono K, Onduka T, Kakuno A, Fujii K (2007) Molecular and histological evaluation of tributyltin toxicity on spermatogenesis in a marine fish, the mummichog (*Fundulus heteroclitus*). Aquat Toxicol 83:73–83

Mochida K, Ito K, Kono K, Onduka T, Kakuno A, Fujii K (2010) Effect of tributyltin oxide exposure on the F-0 and F-1 generations of a marine teleost, the mummichog *Fundulus heteroclitus*. Fish Sci 76:333–341

Morales M, Martinez-Paz P, Ozaez I, Martinez-Guitarte JL, Morcillo G (2013) DNA damage and transcriptional changes induced tributyltin (TBT) after short *in vivo* exposures of *Chironomus riparius* (Diptera) larvae. Comp Biochem Physiol C 158:57–63

Morcillo Y, Janer G, O'Hara SCM, Livingstone DR, Porte C (2004) Interaction of tributyltin with hepatic cytochrome P450 and uridine diphosphate-glucoronosyl transferase systems of fish: in vitro studies. Environ Toxicol Chem 23:990–996

Mortensen AS, Arukwe A (2007) Modulation of xenobiotic biotransformation system and hormonal responses in Atlantic salmon (*Salmo salar*) after exposure to tributyltin (TBT). Comp Biochem Physiol C 145:431–441

Mortensen AS, Arukwe A (2009) Effects of tributyltin (TBT) on *in vitro* hormonal and biotransformation responses in Atlantic salmon (*Salmo salar*). J Toxicol Environ Health A 72:209–218

Morthorst JE, Holbech H, Jeppesen M, Kinnberg KL, Pedersen KL, Bjerregaard P (2014) Evaluation of yolk protein levels as estrogenic biomarker in bivalves; comparison of the alkali-labile phosphate method (ALP) and a species-specific immunoassay (ELISA). Comp Biochem Physiol C 166:88–95

Nakanishi T, Kohroki J, Suzuki S, Ishizaki J, Hiromori Y, Takasuga S, Itoh N, Watanabe Y, Utoguchi N, Tanaka K (2002) Trialkyltin compounds enhance human CG secretion and aromatase activity in human placental choriocarcinoma cells. J Clin Endocrinol Metab 87:2830–2837

Nakanishi T, Nishikawa J, Hiromori Y, Yokoyama H, Koyanagi M, Takasuga S, Ishizaki J, Watanabe M, Isa S, Utoguchi N, Itoh N, Kohno Y, Nishihara T, Tanaka K (2005) Trialkyltin compounds bind retinoid-X receptor to alter human placental endocrine functions. Mol Endocrinol 19:2502–2516

Nakanishi I, Nishikawa J, Tanaka K (2006) Molecular targets of organotin compounds in endocrine disruption: do organotin compounds function as aromatase inhibitors in mammals? Environ Sci 13:89–100.

Nakayama K, Oshima Y, Yamaguchi T, Tsuruda Y, Kang IJ, Kobayashi M, Imada N, Honjo T (2004) Fertilization succes and sexual behavior in male medaka, *Oryzias latipes*, exposed to tributyltin. Chemosphere 55:1331–1337

Nakayama K, Oshima Y, Nagafuchi K, Hano T, Shimasaki Y, Honjo T (2005) Early-life stage toxicity in offspring from exposed parent Medaka, *Oryzias latipes*, to mixtures of tributyltin and polychlorinated biphenyls. Environ Toxicol Chem 24:591–596

Neuwoehner J, Junghans M, Koller M, Escher BI (2008) QSAR analysis and specific endpoints for classifying the physiological modes of action of biocides in synchronous green algae. Aquat Toxicol 90:8–18

Nicolaus EEM, Barry J (2015) Imposex in the dogwhelk (*Nucella lapillus*): 22-year monitoring around England and Wales. Environ Monit Assess 187:736

Nishikawa J, Mamiya S, Kanayama T, Nishikawa T, Shiraishi F, Horiguchi T (2004) Involvement of the retinoid X receptor in the development of imposex caused by organotins in gastropods. Environ Sci Technol 38:6271–6276

Nunez SB, Medin JA, Braissant O, Kemp L, Wahli W, Ozato K, Segars JH (1997) Retinoid X receptor and peroxisome proliferatoractivated receptor activate an estrogen responsive gene independent of the estrogen receptor. Mol Cell Endocrinol 127:27–40

Oberdorster E, McClellan-Green P (2002) Mechanisms of imposex induction in the mud snail, *Ilyanassa obsoleta*: TBT as a neurotoxin and aromatase inhibitor. Mar Environ Res 54:715–718

Oberdorster E, Rittschof D, LeBlanc GA (1998) Alteration of [C-14]-testosterone metabolism after chronic exposure of *Daphnia magna* to tributyltin. Arch Environ Contam Toxicol 34:21–25

OECD (2007) SIDS initial assessment profile for SIAM 24, 19–20 April, 2007. Tributyltin chloride, CAS no. 1461-22-9. http://webnet.oecd.org/Hpv/ui/handler.axd?id=738dc513-4ec8-4d55-ac11-4e1ff79123b8

OECD (2010) Detailed review paper (DRP) on molluscs life-cycle toxicity testing. OECD Environment, Health and Safety Publications, Series on Testing and Assessment, no. 121, pp 182

OECD (2012a) Guidance document on standardised test guidelines for evaluating chemicals for endocrine disruption. Series on testing and assessment, no. 150, pp 524

OECD (2012b) No 142 validation report (phase 2) for the fish sexual development test for the detection of endocrine active substances. Paris, France, OECD ENV/JM/MONO(2011)23/REV1

Oehlmann J, Di Benedetto P, Tillmann M, Duft M, Oetken M, Schulte-Oehlmann U (2007) Endocrine disruption in prosobranch molluscs: evidence and ecological relevance. Ecotoxicology 16:29–43

Ogata R, Omura M, Shimasaki Y, Kubo K, Oshima Y, Aou S, Inoue N (2001) Two-generation reproductive toxicity study of tributyltin chloride in female rats. J Toxicol Environ Health 63:127–144

Ohhira S, Watanabe M, Matsui H (2003) Metabolism of tributyltin and triphenyltin by rat, hamster and human hepatic microsomes. Arch Toxicol 77:138–144

Ohhira S, Enomoto M, Matsui H (2006a) In vitro metabolism of tributyltin and triphenyltin by human cytochrome P-450 isoforms. Toxicology 228:171–177

Ohhira S, Enomoto M, Matsui H (2006b) Sex difference in the principal cytochrome P-450 for tributyltin metabolism in rats. Toxicol Appl Pharmacol 210:32–38

Ohji M, Arai T, Miyazaki N (2003a) Biological effects of tributyltin exposure on the caprellid amphipod, *Caprella danilevskii*. J Mar Biol Assoc UK 83:111–117

Ohji M, Arai T, Miyazaki N (2003b) Chronic effects of tributyltin on the caprellid amphipod *Caprella danilevskii*. Mar Pollut Bull 46:1263–1272

Ohji M, Arai T, Miyazaki N (2006) Transfer of tributyltin from parental female to offspring in the viviparous surfperch *Ditrema temmincki*. Mar Ecol Prog Ser 307:307–310

Omura M, Ogata R, Kubo K, Shimasaki Y, Aou S, Oshima Y, Tanaka A, Hirata M, Makita Y, Inoue N (2001) Two-generation reproductive toxicity study of tributyltin chloride in male rats. Toxicol Sci 64:224–232

Örn S, Yamani S, Norrgren L (2006) Comparison of vitellogenin induction, sex ratio, and gonad morphology between zebrafish and Japanese medaka after exposure to 17α-ethinylestradiol and 17β-trenbolone. Arch Environ Contam Toxicol 51:237–243

Ouadah-Boussouf N, Babin PJ (2016) Pharmacological evaluation of the mechanisms involved in increased adiposity in zebrafish triggered by the environmental contaminant tributyltin. Toxicol Appl Pharmacol 294:32–42

Parrott JL, Bjerregaard P, Brugger KE, Gray Jr LE, Iguchi T, Kadlec SM, Weltje L, Wheeler JR (2017) Uncertainties in biological responses that influence hazard and risk approaches to the regulation of endocrine active substances. Integr Environ Assess Manag 13:293–301

Pascoal S, Carvalho G, Vasieva O, Hughes R, Cossins A, Fang Y-X, Ashelford K, Olohan L, Barroso C, Mendo S, Creer S (2013) Transcriptomics and *in vivo* tests reveal novel mechanisms underlying endocrine disruption in an ecological sentinel, *Nucella lapillus*. Mol Ecol 22:1589–1608

Pavlikova N, Arukwe A (2011) Immune-regulatory transcriptional responses in multiple organs of Atlantic salmon after tributyltin exposure, alone or in combination with forskolin. J Toxicol Environ Health A 74:478–493

Pavlikova N, Kortner TM, Arukwe A (2010) Peroxisome proliferator-activated receptors, estrogenic responses and biotransformation system in the liver of salmon exposed to tributyltin and second messenger activator. Aquat Toxicol 99:176–185

Pereira ML, Eppler E, Thorpe KL, Wheeler JR, Burkhardt-Holm P (2011a) Molecular and cellular effects of chemicals disrupting steroidogenesis during early ovarian development of brown trout (*Salmo trutta fario*). Environ Toxicol 29:199–206

Pereira ML, Wheeler JR, Thorpe KL, Burkhardt-Holm P (2011b) Development of an *ex vivo* brown trout (*Salmo trutta fario*) gonad culture for assessing chemical effects on steroidogenesis. Aquat Toxicol 101:500–511

Rodney SI, Moore DTR (2008) SSD Master, Ver 2.0: determination of hazardous concentrations with species sensitivity distributions. Intrinsik, Toronto, ON, Canada

Saitoh K, Nagai F, Aoki N (2001a) Several environmental pollutants have binding affinities for both androgen receptor and estrogen receptor alpha. J Health Sci 47:495–501

Saitoh M, Yanase T, Morinaga H, Tanabe M, Y-M M, Nishi Y, Nomura M, Okabe T, Goto K, Takayanagi R, Nawata H (2001b) Tributyltin or triphenyltin inhibits aromatase activity in the human granulosa-like tumor cell line KGN. Biochem Biophys Res Commun 289:198–204

Santos MM, Micael J, Carvalho AP, Morabito R, Booy P, Massanisso P, Lamoree M, Reis-Henriques MA (2006) Estrogens counteract the masculinizing effect of tributyltin in zebrafish. Comp Biochem Physiol C 142:151–155

Schlatterer B, Coenen TM, Ebert E, Grau R, Hilbig V, Munk R (1993) Effects of bis(tri-n-butyltin) oxide in Japanese quail exposed during egg laying period: an interlaboratory comparison study. Arch Environ Contam Toxicol 24:440–448

Schwarz TI (2015) The origin of vertebrate steroids in molluscs: uptake, metabolism and depuration studies in the common mussel. PhD thesis, University of Glasgow/Cefas

Schwarz TI, Katsiadaki I, Maskrey BH, Scott AP (2017a) Mussels (*Mytilus* spp.) display an ability for rapid and high capacity uptake of the vertebrate steroid, estradiol-17β from water. J Steroid Biochem Mol Biol 165(B):407–420

Schwarz TI, Katsiadaki I, Maskrey BH, Scott AP (2017b) Rapid uptake, biotransformation, esterification and lack of depuration of testosterone and its metabolites by the common mussel, *Mytilus* spp. J Steroid Biochem Mol Biol 171:54–65

Scott AP (2012) Do mollusks use vertebrate sex steroids as reproductive hormones? Part I: critical appraisal of the evidence for the presence, biosynthesis and uptake of steroids. Steroids 77:1450–1468

Scott AP (2013) Do mollusks use vertebrate sex steroids as reproductive hormones? Part II. Critical review of the evidence that steroids have biological effects. Steroids 78:268–281

Sharan S, Nikhil K, Roy P (2013) Effects of low dose treatment of tributyltin on the regulation of estrogen receptor functions in MCF-7 cells. Toxicol Appl Pharmacol 269:176–186

Sharan S, Nikhil K, Roy P (2014) Disruption of thyroid hormone functions by low dose exposure of tributyltin: an *in vitro* and *in vivo* approach. Gen Comp Endocrinol 206:155–165

Shi H, Zhu P, Guo S (2014) Effects of tributyltin on metamorphosis and gonadal differentiation of *Xenopus laevis* at environmentally relevant concentrations. Toxicol Ind Health 30:297–303

Shimasaki Y, Kitano T, Oshima Y, Inoue S, Imada N, Honjo T (2003) Tributyltin causes masculinization in fish. Environ Toxicol Chem 22:141–144

Shimasaki Y, Oshima Y, Inoue S, Inoue Y, Kang IJ, Nakayama K, Imoto H, Honjo T (2006) Effect of tributyltin on reproduction in Japanese whiting, *Sillago japonica*. Mar Environ Res 62: S245–S248

Si J, Wu X, Wan C, Zeng T, Zhang M, Xie K, Li J (2011) Peripubertal exposure to low doses of tributyltin chloride affects the homeostasis of serum T, E2, LH, and body weight of male mice. Environ Toxicol 26:307–314

Si J, Han X, Zhang F, Xin Q, An L, Li G, Li C (2012) Perinatal exposure to low doses of tributyltin chloride advances puberty and affects patterns of estrous cyclicity in female mice. Environ Toxicol 27:662–670

Si J, Li P, Xin Q, Li X, An L, Li J (2013) Perinatal exposure to low doses of tributyltin chloride reduces sperm count and quality in mice. Environ Toxicol 30:44–52

Silva PV, Silva ARR, Mendo S, Loureiro S (2014) Toxicity of tributyltin (TBT) to terrestrial organisms and its species sensitivity distribution. Sci Total Environ 466–467:1037–1046

Simpson ER, Mahendroo MS, Means GD, Kilgore MW, Corbin CJ, Mendelson CR (1993) Tissue-specific promoters regulate aromatase cytochrome P450 expression. Clin Chem 39:317–324

Smith BS (1981) Male characteristics on female mud snails caused by antifouling bottom paints. J Appl Toxicol 1:22–25

Spence SK, Bryan GW, Gibbs PE, Masters D, Morris L, Hawkins SJ (1990) Effects of TBT contaminations on *Nucella* populations. Funct Ecol 4:425–432

Strand J, Jorgensen A, Tairova Z (2009) TBT pollution and effects in molluscs at US Virgin Islands, Caribbean Sea. Environ Int 35:707–711

Takahashi H (1977) Juvenile hermaphroditism in the zebrafish, *Brachydanio rerio*. Bull Fac Fish Hokkaido Univ 28:57–65

Tian H, Wu P, Wang W, SG R (2015) Disruptions in aromatase expression in the brain, reproductive behavior, and secondary sexual characteristics in male guppies (*Poecilia reticulata*) induced by tributyltin. Aquat Toxicol 162:117–125

Tingaud-Sequeira A, Ouadah N, Babin PJ (2011) Zebrafish obesogenic test: a tool for screening molecules that target adiposity. J Lipid Res 52:1765–1772

Tonk EC, Pennings JL, Piersma AH (2015) An adverse outcome pathway framework for neural tube and axial defects mediated by modulation of retinoic acid homeostasis. Reprod Toxicol 55:104–113

Urishitani H, Katsu Y, Ohta Y, Shiraishi H, Iguchi T, Horiguchi T (2013) Cloning and characterization of the retinoic acid receptor-like protein in the rock shell, *Thais clavigera*. Aquat Toxicol 142-143:203–213

US EPA (1988) Recommendations for and documentation of biological values for use in risk assessment. EPA/600/6-87/008, Feb 1988

US EPA (2003) Ambient aquatic life water quality criteria for Tributyltin (TBT)—final. United States Environmental Protection Agency, Office of Water. EPA 822-R-03-031, Dec 2003

US EPA (2008) Ecological hazard and environmental risk assessment for Tributyltin-containing compounds (TBT). United States Environmental Protection Agency, Office of Prevention, Pesticides, and Toxic Substances, Jan 2008

van Vlaardingen P, Traas TP, Aldenberg T, Wintersen A (2014) ETX 2.1 software. Normal distribution based hazardous concentration and fraction affected. RIVM, The Netherlands

Varella Motta da Costa B, Yogui GT, Souza-Santos LP (2014) Acute toxicity of tributyltin on the marine copepod Tisbe biminiensis. Braz J Oceanogr 62:65–69

Vinggaard AM, Hnida C, Breinholt V, Larsen JC (2000) Screening of selected pesticides for inhibition of CYP19 aromatase activity *in vitro*. Toxicol In Vitro 14:227–234

Vogeler S, Galloway TS, Lyons BP, Bean TP (2014) The nuclear receptor gene family in the Pacific oyster, *Crassostrea gigas*, contains a novel subfamily group. BMC Genomics 15:369

Wang YH, Wang G, LeBlanc GA (2007) Cloning and characterization of the retinoid X receptor from a primitive crustacean *Daphnia magna*. Gen Comp Endocrinol 150:309–318

Yang J, Oshima Y, Sei I, Miyazaki N (2009) Metabolism of tributyltin and triphenyltin by Dall's porpoise hepatic microsomes. Chemosphere 76:1013–1015

Yanik SC, Baker AH, Mann KK, Schlezinger JJ (2011) Organotins are potent activators of $PPAR_{\gamma}$ and adipocyte differentiation in bone marrow multipotent mesenchymal stromal cells. Toxicol Sci 122:476–488

Zhang JL, Zuo ZH, Chen YX, Zhao Y, Hu S, Wang CG (2007) Effect of tributyltin on the development of ovary in female cuvier (*Sebastiscus marmoratus*). Aquat Toxicol 83:174–179

Zhang JL, Zuo ZH, He CY, Cai JL, Wang YQ, Chen YX, Wang CG (2009a) Effect of tributyltin on testicular development in *Sebastiscus marmoratus* and the mechanisms involved. Environ Toxicol Chem 28:1528–1535

Zhang JL, Zuo ZH, He CY, Wu D, Chen YX, Wang CG (2009b) Inhibition of thyroidal status related to depression of testicular development in *Sebastiscus marmoratus* exposed to tributyltin. Aquat Toxicol 94:62–67

Zhang JL, Zuo ZH, Wang YQ, Yu A, Chen YX, Wang CG (2011) Tributyltin chloride results in dorsal curvature in embryo development of *Sebastiscus marmoratus* via apoptosis pathway. Chemosphere 82:437–442

Zhang JL, Zuo ZH, Xiong JL, Sun P, Chen YX, Wang CG (2013a) Tributyltin exposure causes lipotoxicity responses in the ovaries of rockfish, *Sebastiscus marmoratus*. Chemosphere 90:1294–1299

Zhang JL, Zuo ZH, Zhu WW, Sun P, Wang CG (2013b) Sex-different effects of tributyltin on brain aromatase, estrogen receptor and retinoid X receptor gene expression in rockfish (*Sebastiscus marmoratus*). Mar Environ Res 90:113–118

Zhang J, Sun p YF, Kong T, Zhang R (2016) Tributyltin disrupts feeding and energy metabolism in the goldfish (Carassius auratus). Chemosphere 152:221–228

Zuo ZH, Cai JL, Wang XL, Li BW, Wang CG, Chen YX (2009) Acute administration of tributyltin and trimethyltin modulate glutamate and N-methyl-D-aspartate receptor signaling pathway in *Sebastiscus marmoratus*. Aquat Toxicol 92:44–49

Zuo Z, Wu T, Lin M, Zhang S, Yan F, Yang Z, Wang Y, Wang C (2014) Chronic exposure to tributyltin chloride induces pancreatic islet cell apoptosis and disrupts glucose homeostasis in male mice. Environ Sci Technol 48:5179–5186

Oxidative Stress and Heavy Metals in Plants

Radka Fryzova, Miroslav Pohanka, Pavla Martinkova, Hana Cihlarova, Martin Brtnicky, Jan Hladky, and Jindrich Kynicky

Contents

R. Fryzova • M. Brtnicky • J. Hladky • J. Kynicky (✉)
Department of Geology and Pedology, Faculty of Forestry and Wood Technology, Mendel University in Brno, Zemedelska 3, Brno 613 00, Czech Republic

Central European Institute of Technology, Brno University of Technology, Purkynova 123, Brno 612 00, Czech Republic
e-mail: fryzova@post.cz; martin.brtnicky@seznam.cz; pudoznalec@gmail.com; jindrak@email.cz

M. Pohanka
Department of Geology and Pedology, Faculty of Forestry and Wood Technology, Mendel University in Brno, Zemedelska 3, Brno 613 00, Czech Republic

Faculty of Military Health Sciences, University of Defence, Trebesska 1575, Hradec Kralove 500 01, Czech Republic
e-mail: miroslav.pohanka@gmail.com

P. Martinkova
Central European Institute of Technology, Brno University of Technology, Purkynova 123, Brno 612 00, Czech Republic

Faculty of Military Health Sciences, University of Defence, Trebesska 1575, Hradec Kralove 500 01, Czech Republic
e-mail: pavla.martinkova@unob.cz

H. Cihlarova
Department of Geology and Pedology, Faculty of Forestry and Wood Technology, Mendel University in Brno, Zemedelska 3, Brno 613 00, Czech Republic
e-mail: hana.cihlarova@mendelu.cz

P. de Voogt (ed.), *Reviews of Environmental Contamination and Toxicology Volume 245*, Reviews of Environmental Contamination and Toxicology 245,
DOI 10.1007/398_2017_7

1 Introduction

In the nature, heavy metals induce serious contamination because of their persistence, high toxicity, and easy transmission through the food chain. It needs to be emphasized that there is no agreement as to what metals cannot be considered heavy. Currently, metals with molecular weight over 20 g/mol and/or density higher than 5 g/mL are designated as heavy metals (Duffus 2002, 2003; Stankovic et al. 2014; Kim et al. 2015).

Cd, Hg, Pb, and Tl are typical metals that meet the definition and fall into the group of heavy metals because of their density and molecular weight. These metals can be found in the nature; however, they have no biological role in animals nor plants. Other metals such as Co, Cu, Fe, Mn, and Zn also meet the definition of heavy metals and they are necessary for keeping homeostasis until their level does not exceed threshold level for toxicity (Van Bussel et al. 2014). Some of the metals are also typical radioisotopes (Pu and U, for instance), hence this fact should be taken into consideration and there should be considered whether the toxicity is mediated by the metal properties or there is a pathological process caused by exposure to ionizing radiation.

Harmful effect of heavy metals is not an easy effect based on an interruption of a single pathway or interaction with a single molecule. On the contrary, heavy metals have impact on multiple processes in the body, resulting in pathological consequences. In humans, association between heavy metals and some neurodegenerative disorders like Alzheimer disease, kidney damage, and cancer are discussed (Pohanka 2014a, b). Apart from the pathologies, heavy metals can be stored in cells. Both plants and microbes have high potential to accumulate heavy metals, which is useful in bioremediation (Wang and Sun 2013; Topolska et al. 2014; Hechmi et al. 2015). On the other hand, heavy metals stored in plants may have toxic effect on animal and human organisms when taken through the food chain.

This review is focused on summarizing the role of heavy metals in plants, pathology mechanisms, and environmental risks caused by the metals. The motivation for writing this paper was triggered by the fact that despite significant recent progress in the research of heavy metals, the issue has not been extensively reviewed. A complex overview of the issue is given in this manuscript, and the impact of heavy metal toxicity, accumulation in plants, chemical assays and the link between heavy metals and oxidative stress in plants are also discussed here.

2 Mobility of Heavy Metals

Metal mobility is influenced by several factors and soil properties such as the content of organic matter, oxides as well as soil structure and profile development (Mehes-Smith et al. 2013a). Transport of metals may be enhanced by several factors including metal association with mobile colloidal size and formation of metal organic and inorganic complexes that do not sorb to soil solid surfaces (Puls et al. 1991). According to Mehes-Smith et al. (2013a), soil topography plays a key role in metal distribution and horizontal mobility. Metals are mobilized by being captured by root cells from soil particles, bound by the cell wall and then transferred across the plasma membrane, driven by ATP-dependent proton pumps that catalyze H^+ extrusion across the membrane (Singh et al. 2003). On binding, they displace cations such as Ca^{2+} and Mg^{2+} from the cell walls and membranes (Brunner et al. 2008; Flouty and Khalaf 2015). Binding studies with pectins have demonstrated that the binding preferences are: $Al^{3+} > Cu^{2+} > Pb^{2+} > Zn^{2+} = Ca^{2+}$ or $Cu^{2+} = Pb^{2+} > Cd^{2+} = Zn^{2} > Ca^{2+}$ (Franco et al. 2004), $Pb^{2+} >> Cu^{2+} > Co^{2+} > Ni^{2+} >> Zn^{2+} > Cd^{2+}$ (Kartel et al. 1999) or $Pb^{2+} = Cd^{2+}$ (Debbaudt et al. 2004), depending on the origins of the pectin. Pectins are a family of heterogeneous polysaccharides present in the primary cell wall and in the middle lamella of plant tissue (Guo et al. 2015). In enriched heavy metal environments, some plants will elevate the capacity of their cell wall to bind metals by increasing polysaccharides, such as pectins (Colzi et al. 2011). However, the binding capacity can be altered also for other reasons. The main functional groups of pectin are: hydroxyl, carboxyl, amide, and methoxyl, and these have been traditionally associated with heavy metal binding, especially carboxyl groups (which enable the binding of divalent and trivalent heavy metals ions) with great biosorption and heavy metal removal potential (Mata et al. 2009). The quantitative data on adsorption performance of pectins suggest their applicability as food additives or remedies for efficient removal of Pb^{2+}, Cu^{2+}, Co^{2+}, and Ni^{2+} ions from different biological systems, including human and animal organisms (Kartel et al. 1999).

3 Bioavailability and Bioaccessibility

The terms bioaccessibility and bioavailability mean that the chemicals (metals) can interact with the organism and became available in the parts of organism and can take an effect. Assessment of the levels of metal bioavailability and bioaccessibility is critical in understanding the possible effect on soil biota (Ettler et al. 2012). In general, the availability of metals for plants depends on soil pH and on organic matter contents (De Matos et al. 2001; Stankovic et al. 2014), yet there are no standardized protocols for estimating the bioavailable metal content in soil. However, root-colonizing bacteria and mycorrhiza can significantly increase the bioavailability of various heavy metal ions for uptake (Singh et al. 2003). The

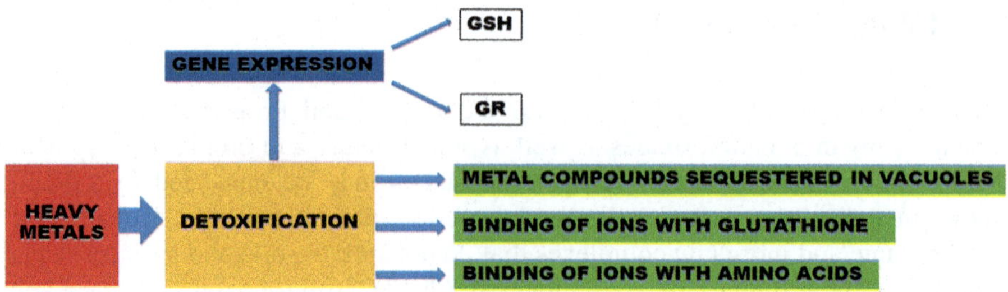

Fig. 1 Survey of heavy metals interaction with plant body

mobility, bioavailability, and potential toxicity of a metal in the soil depend on its concentration in a soil solution, the nature of its association with other soluble species and the soil's ability to release the metal from the solid phase to be acquired by the plants (Violante et al. 2010). Plants with a high bioconcentration factor (BCF—metal concentration ratio of plant roots to soil) and low translocation factor (TF—ratio of the concentration of the total amount of metal in leaves compared to the metal content in roots), e.g. *Deschampsia cespitosa*, have the potential for stabilization of ecological system (Galfati et al. 2011). Although tree fine roots adapt well to conditions with heavy metal contamination, their phytostabilization capabilities seem to be very low (Brunner et al. 2008; Kalubi et al. 2016). A survey of heavy metals interaction with a plant body is depicted in Fig. 1.

4 Heavy Metals in Plants

Heavy metals are considered to be an important stress factor for plants. Pathways of heavy metals in plants are briefly summarized in Fig. 2. If a natural amount of heavy metals is present in soils, plants are able to avoid their negative impact (Juknys et al. 2012). However, high concentrations of heavy metals cause harmful effect on cellular and physiological processes in plants (Ma 2005; Dimkpa et al. 2009). The most abundant heavy metals in soils are Fe and Al. Fe is one of essential elements and plays an important role in plant nutrition. Mo and Mn are also important micronutrients: Zn, Ni, Cu, V, Co, W, and Cr belong among trace elements and As, Hg, Ag, Sb, Cd, Pb have no nutritional or physiological function in plants (Schützendübel and Polle 2001). Negative impact of these elements is highly dependent on their concentration in soils as well as in plants (Stankovic et al. 2014). Uptake of heavy metals is influenced by abiotic and biotic conditions. The value of pH is crucial for the solubility of heavy metals, for example Al and Fe become soluble and toxic in pH under 7. Another important parameter of heavy metal uptake are the processes of adsorption, desorption, and complexation, which are associated with individual soil types. Generally, there are two ways of heavy metal uptake: passive uptake, based on the concentration gradient across the membrane and inducible substrate-specific and energy-dependent uptake. Heavy metals

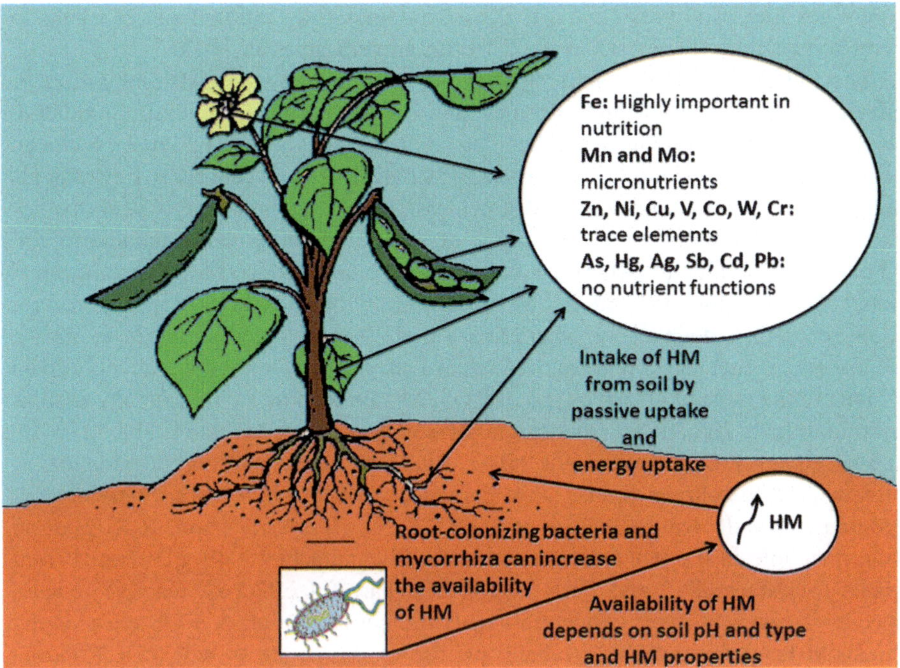

Fig. 2 Availability and utilization of heavy metals (HM) in plants

uptake from soil to roots and the subsequent behavior of heavy metals on the level of tissue and cells are very complex. However, it is necessary to understand this mechanism to be able to use it in phytoremediation and improve plant tolerance to heavy metals (Schützendübel and Polle 2001). The issue remains unresolved despite great effort which has been made since the introduction of phytoremediation. Still, relevant genes were determined and recent data seem to be promising for future applications (Ahmad et al. 2016). *Lolium perenne* (Rees et al. 2015), *Alyssum montanum* (Agrawal et al. 2013), *Stanleya albescens* (Freeman et al. 2010), *Nicotiana tabacum* and *Brassica juncea* (Liang et al. 2009), *Senecio coronatus* (Mesjasz-Przybyłowicz et al. 2007), *Thlaspi arvense* and *Thlaspi perfoliatum* (Freeman et al. 2005) can be mentioned as the typical examples of promising plants (see further text).

There are several strategies plants employ to cope with high concentrations of heavy metals in soils. The first method is the prevention of the intake of heavy metals into plant roots. Heavy metals can be removed from plants by mycorrhizal association, metal sequestration, or complexation by exuding organic compounds from roots (Verbruggen et al. 2009; Antosiewicz et al. 2014). These mechanisms usually immobilize heavy metals. If the plant is not able to avoid heavy metal uptake, tolerance mechanisms for detoxification are activated. These are metal sequestration and compartmentalization in various intracellular compartments, metal ions trafficking, metal binding to cell wall, biosynthesis or accumulation of osmolytes and osmoprotectants. If these strategies fail, antioxidants can help to keep homeostasis (Gajewska et al. 2013; Emamverdian et al. 2015).

There are several phytotoxicological impacts of heavy metals: inhibition of enzymes, inactivation of biomolecules, and oxidative stress. Generally, oxidative stress is defined as imbalance between pro-oxidant and antioxidant level which can be also entitled as oxidative homeostasis. Oxidative stress is caused by free radicals and molecules containing activated atoms of oxygen called reactive oxygen species (ROS), and it is associated with a lack of electrons which cause damage to cell compounds (Demidchik 2015). Oxidative stress can be induced by high concentrations of heavy metals, and it causes inhibition of growth. There are several ROS-generating mechanisms where heavy metals participate (Juknys et al. 2012). Redox-active metals (copper and chrome) can produce hydroxyl radicals (OH•) which are the most aggressive type of ROS. Metals without redox capacity such as cadmium, lead, zinc, nickel, etc. can involve singlet oxygen which is able to create another type of ROS superoxide (O_2•). ROS can cause unspecific oxidation of proteins and membrane lipids, DNA damage, enzyme inhibition by activation of programmed cell death (Sharma et al. 2012). On the other hand, ROS play an important role in the plant defense system and cannot be totally eliminated from plants (Schützendübel and Polle 2001). The harmful effect of reactive oxygen species is given by their concentration. If the concentration of ROS exceeds the threshold level for defense mechanisms, the oxidative stress will arise (Sharma et al. 2012).

5 Toxicity of Heavy Metals

Soils contaminated with heavy metals represent an escalating problem for all living organisms, such as plants, animals, or humans. In general, mutagenic ability of heavy metals (Knasmüller et al. 1998) and oxidative stress induction via Fenton and Haber-Weiss reactions can be mentioned (Jomova and Valko 2011) and many others can be learned from literature. Adaptive responses of plants to heavy metal-contaminated environments are efficient processes including many physiological, molecular, genetic, and ecological traits, which give certain species the ability to survive or to hyperaccumulate the toxic metals (Sarma 2011).

6 Heavy Metal Phytotoxicity and Oxidative Stress Arise

Heavy metal phytotoxicity may result from alterations of numerous physiological processes caused at cellular/molecular level by inactivating enzymes, blocking functional groups of metabolically important molecules, displacing or substituting for essential elements and disrupting membrane integrity, and the phytotoxicity is an addition of reactive oxygen or nitrogen species releasing in natural pathways like

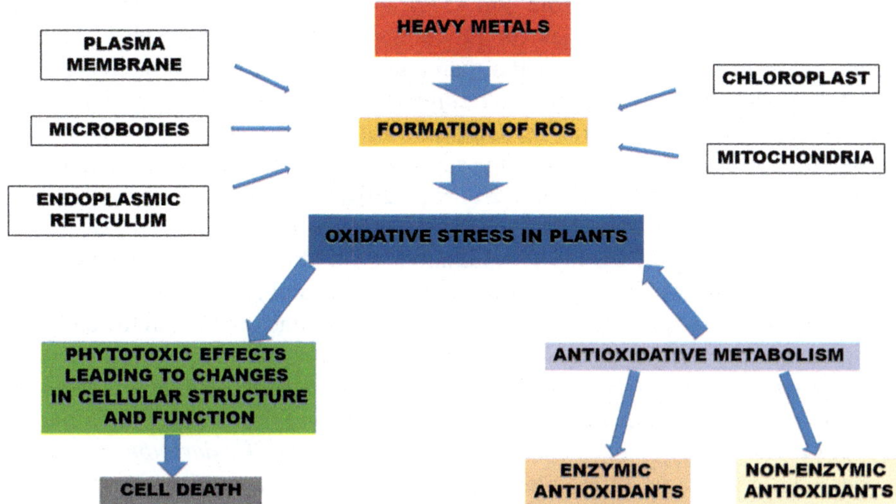

Fig. 3 Link between oxidative stress and heavy metals in plants

photosynthesis, tricarboxylic acid cycle, and Calvin cycle (Rascio and Navari-Izzo 2011). Fenton reaction can be another pathway where both antioxidants and metals are necessary and interact together. Here, reactive species are results of the interaction. The processes are summarized in Fig. 3. Metal-induced oxidative stress has been strongly linked to early toxicity symptoms (Sharma and Dietz 2009). Complexation with metal-binding peptides, metallothioneins (MTs, gene-encoded polypeptides), and phytochelatins (PCs, enzymatically synthesized peptides) results in formation of non-toxic metabolites protecting cells and the whole organism (Singh et al. 2003). MTs have three characteristic motifs based on their cysteine content and structure: Cysteine-Cysteine, Cysteine-X-Cysteine, and Cysteine-X-X-Cysteine, in which X denotes any amino acid (Mejáre and Bülow 2001). In angiosperms, MTs can be classified into four types based on the conserved positions of Cys residues (Cobbet and Goldsbrough 2002), with some general trends in the expression: type-1 MTs expressed predominantly in roots, type-2 MTs in leaves, type-3 MTs in fruits, and type-4 MTs in seeds (Guo et al. 2008). PCs are tricarboxylic sometimes confusingly described as class III MTs (Cobbet and Goldsbrough 2002) and can be divided into five main classes: canonical PCs, homo-PCs [*iso*(PC)(β-Ala)], hydroxymethyl-PCs [*iso*(PC)(Ser)], iso-PCs [*iso*-PC (Glu)], and desglycine-PCs [*des*(Gly)PC], containing n γ-Glutamyl-Cysteine repeats capped C-terminally by a Glycine, β-Alanine, Serine, Glutamic acid, or no residue, respectively (Zenk 1996).

Metal hyperaccumulation typically occurs in over 500 plant species (yet many are still unidentified) and approximately in 0.2% of all angiosperms (Sarma 2011), and metal accumulating plants (of soils in both tropical and temperate zones of all the continents) are model plants for phytoremediation (use of plants to ameliorate contaminated sites) and phytomining. The ratio of metals between soil and plant parts—metal transfer factors from soil (TFS)—is an important criterion for the selection of model plant species for phytoremediation; ratio >1 means higher accumulation of metals in plant parts than soil (Barman et al. 2000). According to Kumar et al. (2013), *Parthenium hysterophorus* and *Spinacia oleracea* have TFS above 1 for Cr, Cu, Ni, Pb, and Cd. *Impatiens walleriana* is able to accumulate a tenfold higher concentration of Cd in the shoot than a typical hyperaccumulator and its TFS values are greater than unity (Wei et al. 2012), while *Noccaea caerulescens* has TFS >1 for Ni as a consequence of inoculation (Visioli et al. 2014). Farahat and Linderholm (2015) stated that the accumulation of Zn, Mn, Cu, and Cd with transfer factors >1 for wastewater-irrigated trees indicated the ability for metal accumulation of *Cupressus sempervirens*. Sainger et al. (2011) found out that on the basis of TFS greater than 1, eight plant species for Zn (*Achyranthes aspera*, *Amaranthus viridis*, *Senna occidentalis*, *Chenopodium album*, *Croton bonplandianum*, *Cynodon dactylon*, *Saccharum munja*, *Tephrosia purpurea*) and Fe (*Vachellia nilotica*, *A. aspera*, *A. viridis*, *S. occidentalis*, *C. bonplandianum*, *C. dactylon*, *Physalis minima*, *T. purpurea*), three plant species for Cu (*A. aspera*, *P. minima*, *S. munja*), and two plant species for Ni (*P. minima*, *T. purpurea*) could be considered as hyperaccumulators (HAs) and used in phytoextraction technology. The aforementioned HAs are summarized in Table 1.

HAs are plants that belong to distantly related families, but share the ability to grow on metalliferous soils. These plants are able to accumulate extraordinarily high amounts of heavy metals in the aerial organs, far in excess of the levels found in the majority of species, without suffering phytotoxic effects (Rascio and Navari-Izzo 2011), respectively possesses genetically inherited traits of metals hyperaccumulation and tolerance. These plants are differentiated from non-hyperaccumulating species (NHAs), e.g. *Lolium perenne* (Rees et al. 2015), *Alyssum montanum* (Agrawal et al. 2013), *Stanleya albescens* (Freeman et al. 2010), *Nicotiana tabacum* and *Brassica juncea* (Liang et al. 2009), *Senecio coronatus* (Mesjasz-Przybyłowicz et al. 2007), *Thlaspi arvense* and *Thlaspi perfoliatum* (Freeman et al. 2005), by a highly increased ratio of heavy metal uptake (at concentrations 100- to 1000-fold higher than those found in NHAs). An important role is also played by faster translocation of metals from roots to shoots, and a better ability to detoxify and sequester heavy metals in leaves, where constitutive overexpression of genes, e.g. *Os*ZIP4 and *Ah*HMA4 (Antosiewicz et al. 2014), *Cs*HMA3 (Park et al. 2014), *At*HMA4 (Verret et al. 2004; Siemianowski et al. 2014), *Sa*MTP1 (Zhang et al. 2011), *Tc*YSL3 (Gendre et al. 2006), encodes transmembrane transporters. Ricachenevsky et al. (2013) revealed that although useful in basic studies of gene function, constitutive over-expression changes metal homeostasis in all organs, and often affects the uptake and distribution of more than one element.

Mechanisms involved in heavy metal increased tolerance and heavy metal distribution in an excluder non-hyperaccumulator and a hyperaccumulator plant: heavy metal binding to the cell walls and/or cell exudates, root uptake, chelation in the cytosol and/or sequestration in vacuoles, root-to-shoot translocation. The spots indicate the plant organ in which the different mechanisms occur and the spot sizes

Table 1 Selected plants (hyperaccumulators) with high efficacy to accumulate heavy metals

Metal	Plant	Family
Zn	*Achyranthus aspera*	Amaranthaceae
	Amaranthus viridis	Amaranthaceae
	Senna (Cassia) occidentalis	Fabaceae
	Chenopodium album	Amaranthaceae
	Croton bonplandianum	Euphorbiaceae
	Cyanodon dactylon	Poaceae
	Saccharum munja	Poaceae
	Tephrosia purpurea	Fabaceae
Fe	*Vachellia (Acacia) nilotica*	Fabaceae
	Achyranthus aspera	Amaranthaceae
	Amaranthus viridis	Amaranthaceae
	Senna (Cassia) occidentalis	Fabaceae
	Croton bonplandianum	Euphorbiaceae
	Cyanodon dactylon	Poaceae
	Physalis minima	Solanaceae
	Tephrosia purpurea	Fabaceae
Cu	*Achyranthus aspera*	Amaranthaceae
	Physalis minima	Solanaceae
	Saccharum munja	Poaceae
Ni	*Saccharum munja*	Poaceae
	Tephrosia purpurea	Fabaceae
Cd	*Cupressus sempervirens*	Cupressaceae

Data acquired from Sainger et al. (2011), Farahat and Linderholm (2015)

indicate their level. According to the elemental defense hypothesis the high heavy metal concentrations make hyperaccumulator leaves poisonous to herbivores (Rascio and Navari-Izzo 2011).

Initially, the term HAs referred to plants that were able to accumulate more than 1 mg/g of Ni (dry weight) in the shoot, an exceptionally high heavy metal concentration considering that in vegetative organs of most plants Ni toxicity starts from 10 to 15 μg/g (Rascio and Navari-Izzo 2011). As defined by specific phytotoxicity of other heavy metals—HAs are plants that, when growing on native soils, concentrate >10 mg/g (1%) Mn or Zn, >1 mg/g (0.1%) As, Co, Cr, Cu, Pb, Sb, Se or Tl, and >0.1 mg/g (0.01%) Cd in the aerial organs, without suffering phytotoxic damage (Verbruggen et al. 2009). Ni is hyperaccumulated by the largest amount of species (more than 75%) and reaches the highest concentration in a plant, while a low number of HAs have been found for Cd, which is one of the most toxic heavy metals (Rascio and Navari-Izzo 2011). Zn HAs are likewise less numerous. For example, the *Thlaspi* family are hyperaccumulating plants among which 23 species hyperaccumulate Ni, 10 species hyperaccumulate Zn, just 3 species (*T. caerulescens, T. praecox,* and *T. goesingense*) hyperaccumulate Cd and 1 species (*T. praecox)* hyperaccumulates Pb (Miransari 2011).

To be considered as HAs, plants have to meet also these criteria: shoot/root quotient higher than 1, i.e. level of heavy metal in the shoot divided by the level of heavy metal in the root, and extraction coefficient higher than 1, i.e. level of heavy metal in the shoot divided by the total level of heavy metal in the soil (Mganga et al. 2011).

Plants growing on metalliferous soils can be grouped into three categories: excluders, indicators, and accumulators/HAs (Bhargava et al. 2012). Most HAs are endemic to metalliferous soils, behaving as "strict metallophytes," whereas some "facultative metallophytes" can live also on non-metalliferous ones, although they are more prevalent on metal-enriched habitats (Assunção et al. 2003). Moreover, there are species that embrace both metallicolous and non-metallicolous populations.

An interesting feature revealed by research is that most key steps in hyperaccumulation do not rely on novel genes, but depend on genes common to HAs and NHAs that are differently expressed and regulated in the two kinds of plants (Verbruggen et al. 2009), such as members of ZIP (Zinc-regulated transporter, Iron-regulated transporter Protein), HMA (Heavy Metal transporting ATPase), MATE (Multidrug And Toxin Efflux), YSL (Yellow Strip 1-Like protein), CAX (Cation Exchanger), NRAMP (Natural Resistance-Associated Macrophage Protein), and CDF (Cation Diffusion Facilitator), resp. MTP (Metal Tolerance Protein) families. Lin et al. (2014) listed in their study also genes belonging to PCR (Plant Cadmium Resistance), PDR (Pleiotropic Drug Resistance protein), and PDF (Plant Defensin) families, as well as transcripts related to metal chelator and metal chelator transporter functions, such as genes of NAS (Nicotianamine Synthase), PCS (Phytochelatin Synthase), MT, ZIF (Zinc Induced Facilitator), and MRP/ABC (Multidrug Resistance-associated Protein/ATP-Binding Cassette transporter). De Abreu-Neto et al. (2013) studied two large families of genes encoding HIPP (Heavy metal-associated Isoprenylated Plant Protein) and HPP (Heavy metal-associated Plant Protein). In addition, Krämer et al. (2007) reported other well-known proteins that mediate the transport of transition metals in plants, which belong to the following families: OPT (Oligopeptide Transporter), MFS (Major Facilitator Superfamily), COPT (Copper Transporter), CCC1 (Ca^{2+}-sensitive Cross Complementer 1), IREG (Iron-Regulated protein) and ATM (ABC Transporters of the Mitochondria). Several cation transporters have been identified in recent years, most of which are also in the SAMS (S-Adenosyl-Methionine Synthetase) and FER (Ferritin Fe (III) binding) families (Bhargava et al. 2012). Yet not every member of all these families listed above has been functionally characterized and therefore, further study is required before their roles and functions are fully understood.

Molecular technique provides better understanding of the gene regulation systems and plant metal homeostasis. In order to protect themselves from the oxidative stress, plants have several antioxidative defense systems to scavenge toxic radicals. This defensibility is divided into two main classes: low molecular weight antioxidants, which comprise lipid-soluble membrane-associated antioxidants (e.g., α-tocopherol, β-carotene) and the water-soluble reductants (e.g., glutathione, ascorbate); and antioxidative enzymes—superoxide dismutase (SOD), ascorbate peroxidase (APX), catalase (CAT), and glutathione reductase (GR) (Sarma 2011; Zaimoglu et al. 2011; Gupta and Ahmad 2013; Mehes-Smith et al. 2013b; Adrees et al. 2015; Lou et al. 2015).

There are three main strategies involved during the detoxification of metallic ions: phytovolatilization and/or chemical transformation (the chemical conversion of toxic elements into less toxic and volatile compounds results in the removal of harmful elements from plant tissues), efflux from the cytoplasm, and binding or chelation of trace elements (Singh et al. 2003). In particular, chelation is the most widespread intracellular mechanism for the maintenance of low concentrations and detoxification of free ion metals in plant cytoplasm that can be performed by thiol compounds (such as tripeptide glutathione, metallothioneins (MTs), phytochelatins), and also by non-thiol compounds (such as organic acids and amino acids, e.g. histidine, nicotianamine) (Seth et al. 2012; Anjum et al. 2015). GSH, a tripeptide (γ-Glu-Cys-Gly), is recognized as an antioxidant that plays a key role in the defense mechanism of plants (Nahar et al. 2015) and it is a precursor for the synthesis of phytochelatins (family of peptides structurally related to GSH) in metal-exposed plants (Hossain and Komatsu 2013). MTs are low-molecular-weight, cysteine-rich proteins that have high affinity for binding metal cations, and their overexpression can increase plant tolerance to specific metals (Fernandez et al. 2012; Lv et al. 2013). Other compounds like the phenolic one can be involved in the protection and compounds like anthraquinones and flavonoids are important. Phytochelatins (PCs) are also a family of metal-complexing peptides which are rapidly induced on overexposure to metals or metalloids in plants, animals, and some yeasts (Vatamaniuk et al. 2001) and they can bind metals possessing a high affinity to sulfhydryl groups (Anjum et al. 2015).

7 Heavy Metals and Plant Saccharide Metabolism

7.1 Plant Saccharide Metabolism

Plant saccharide metabolism depends on photosynthesis and the Calvin cycle, i.e. two elemental processes that take place in a plant cell, namely inside the chloroplast. Photosynthesis is a process where water and sunlight are absorbed from the air and converted into chemical macroergic compounds (ATP, NADPH) used in subsequent reactions for production of proteins, fats, and carbohydrates. The Calvin cycle is a series of reactions following photosynthesis, where produced ATP and NADPH are utilized for the formation of 3-phosphoglyceraldehyde from carbon dioxide and ribulose-1,5-bisphosphate. Subsequently, D-fructose-6-phosphate as a precursor of D-glucose-6-phosphate and of D-mannose-6-phosphate is enzymatically produced (Fig. 4). These basic saccharides can be converted into other saccharides or can enter protein, fat, or nucleic acid metabolism (Velisek and Cejpek 2005a; Tamoi et al. 2006).

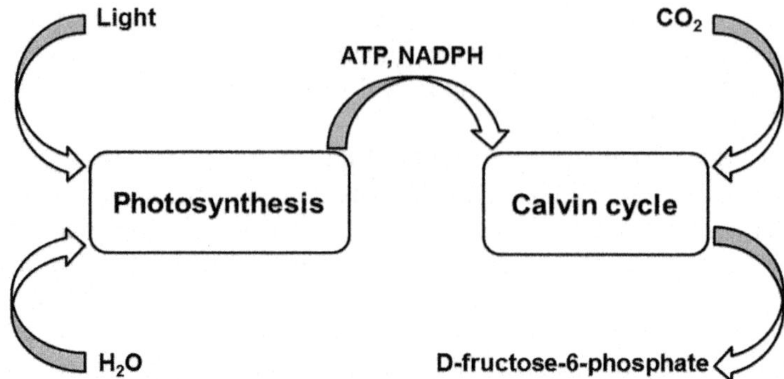

Fig. 4 Scheme of basic plant metabolism

Saccharides, of both plant and animal origin, are the primary products of photosynthesis and the Calvin cycle and they can be divided into three groups: monosaccharides, oligosaccharides, and polysaccharides. Monosaccharides are the most basic and the most important part of plant metabolism. Types of monosaccharides are distinguished according to the count of carbon molecules in their structure; five- and six-carbon sugars belong to the most frequently occurring ones in the nature. 3-phosphoglyceraldehyde produced during the first enzymatically catalyzed reaction of the Calvin cycle belongs to three-carbon sugars, and its two molecules are important for the synthesis of D-fructose-6-phosphate, the precursor of all plant sugars. Subsequent monosaccharide metabolic processes are shown in Fig. 5 (Velisek and Cejpek 2005a; Hu et al. 2014).

Oligosaccharides can be defined as low-molecular-weight carbohydrates consisting of monosaccharide units (at least two but no more than ten of them) held together by a glycosidic bond. Polysaccharides are macromolecules of carbohydrates consisting of monosaccharide units (more than ten) that are also held together by a glycosidic bond. The molecular weight of oligosaccharides ranges between that of monosaccharides and polysaccharides in the nature. All types of sugars have great meaning for the plant itself as well as for animal and human diet and development, structure, and metabolism of these organisms (Velisek and Cejpek 2005b; Mussatto and Mancilha 2007; Kurd and Samavati 2015).

7.2 *Important Plant Saccharides*

Glucose and fructose are the most important monosaccharides in plant metabolism. They have a storage function (preservation of energy as monosaccharides or as oligosaccharides) and a structural function (structural polysaccharides such as starch and cellulose) (Velisek and Cejpek 2005b; Hieu et al. 2015). Fructose enters glycolysis and respiration and it also serves as a precursor for fat and protein synthesis. Fructose can also be transformed into activated forms of glucose and

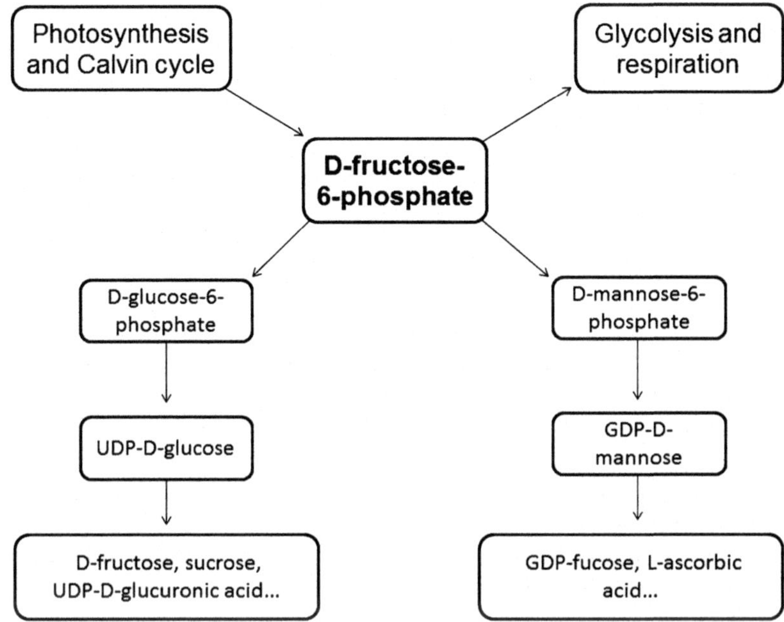

Fig. 5 Scheme of plant saccharide metabolism

mannose, precursors of the synthesis of other mono-, oligo-, and polysaccharides (Velisek and Cejpek 2005a). Glucose and fructose have the function of signal molecules; they are able to begin hexose-based metabolism in the membrane or in the cytoplasm (Winter and Huber 2000; Koch 2004). The activated form of glucose (UDP-D-glucose) also serves as a precursor for the synthesis of UDP-D-glucuronic acid, and this conversion is the first step of the synthesis of nucleotide sugars (UDP-D-apiose, UDP-L-arabinose, and UDP-D-xylose) (Reiter and Vanzin 2001).

Sucrose and trehalose belong to one of the few free and the most frequently naturally occurring non-reducing disaccharides in plants. Unlike trehalose, which can only be found in few higher plant species, sucrose occurs even in algae, cyanobacteria, lower plants and all oxygenic photosynthetic organisms (Wingler 2002; Salerno and Curatti 2003). Sucrose, which consists of fructose and glucose units held together by a glycosidic bond, has a major role in many processes like cell growth, development, signaling, etc. (Cumino et al. 2002; Hieu et al. 2015; Rorabaugh et al. 2015). It has an important metabolic role as a donor of glucosyl and fructosyl moiety initiating the synthesis of polysaccharides and nucleotide sugars. Trehalose is composed of two glucose units connected by a glycosidic bond and it has storage, transport, and protective functions during heat stress and dehydration (Wingler 2002; Salerno and Curatti 2003).

Cellulose, callose, and pectin are the most important plant cell wall polysaccharides. Starch and fructans are the major plant storage sugars (Velisek and Cejpek

2005b). Both cellulose, i.e. a homogenous polymer of β-(1,4) glucose, and callose, i.e. a β-(1,3) glucose polymer, are synthesized from UDP-glucose, but cellulose is generated in the plasma membrane, and callose-like pectin originates in the Golgi apparatus (Amor et al. 1995; Perrin 2001; Reiter 2002). Unlike callose and starch, glucose residues of cellulose are repeated in their structural chains (Brown et al. 1996). Cellulose is responsible for cell growth, shape, tissue morphology, and extension (Perrin 2001). Callose is rarely generated within *in vivo* plant metabolism; however, it can be found in cell plates, sieve tubes and during some stages of mega- and microsporogenesis. Callose is responsible for plants' wound, stress and infection responses, and it is part of structures involved in cell growth and differentiation (Brown et al. 1996; Kudlicka and Brown 1997). Pectin is composed of α-1,4-glucuronic acid residues occurring in the primary wall of all higher plants except grasses and relative plants, and it plays a major role in cell wall structure, growth, and development (Velisek and Cejpek 2005b).

Starch is a plant polysaccharide consisting of two glucose homopolymers—amylose and amylopectin. Amylopectin is structurally close to the animal polysaccharide of glycogen, it is highly branched and formed of α-(1,4)-linked and α-(1,6)-branched glucose chains (Myers et al. 2000; Li et al. 2015a, b). Amylose has a linear structure consisting of glucose units with α-(1,4) bonds. Starch is generated in leaves during the day when it is stored in chloroplasts, and it is utilized for sucrose synthesis during the night. Fructans serve for the deposition of energy in flowering plants. They are made up of β-(1,2) fructose units linked together, and the end unit of glucose is usually bound (Velisek and Cejpek 2005b; Wasserman et al. 2015).

7.3 Heavy Metal Toxic Effect on Plant Saccharides

Saccharides biosynthesis pathway in plant metabolism includes fixation of carbon dioxide (the Calvin cycle), metabolic transformation of basic monosaccharides into other monosaccharides, and finally the distribution of saccharides throughout the plant and their storage or utilization (Slewinski and Braun 2010). Metals, such as Zn, Fe, Cu, and others, have great influence on plant metabolism and structure and they are required as co-factors in many enzymatically catalyzed reactions. Their major contribution is based on the ability to affect plant photosynthesis, respiration, ethylene perception, circadian clock, and programmed cell death, and they participate on the protection of plant metabolism (Clemens 2006; Garcia et al. 2014). On the other hand, higher concentrations of metals are toxic for plants. Most metals inhibit plant growth due to their interference with elements that are essential for enzymatic functions. Additionally, natural redox activity of heavy metals (Cu, Fe) requested under physiological conditions may increase the plant's production of reactive oxygen species in high metal concentrations. The synthesis of chlorophyll and other photosynthetic pigments in plant leaves is also decreased by the long-term effect of heavy metals. Reduction of pigments is caused either due to the inhibition of the enzymatic system of chloroplasts or due to insufficiency of essential compounds (Das et al. 1997; Chettri et al. 1998; Fargasova 2004; Clemens 2006; Mera et al. 2016). As a result of human activity, heavy metals begin to be more available in soils and natural sources, ecosystems become damaged, and plant metabolism is disrupted (Das et al. 1997; He et al. 2015).

The toxic effect of heavy metals also leads to inhibition of saccharide metabolism in plants. Four basic metals have been reported as toxic for plant metabolism. Ni is able to affect the mobilization of saccharides in germinating seeds. Pb may decelerate root growth via increased saccharide contents and retard saccharide metabolizing enzymes: α-amylase, β-amylase and invertase as well as Cu and Cd (Solanki and Dhankhar 2011; Sethy and Ghosh 2013).

The effect of Cd on plant metabolism was examined in many studies. Cd is toxic for plants where it becomes accumulated and also for animals that eat plants affected by this metal. Nevertheless, Cd has the most toxic effect on the human body where it gets accumulated in organs after the consumption of food containing Cd (Kirkham 2006).

Hédiji et al. (2010) studied the long-term effect of Cd on the growth and metabolic profile of tomato plants, namely its influence on the metabolism of carotenoids, carbohydrates, organic acids, and amino acids. Tomato plants were treated with two concentrations of Cd (20 and 100 μmol/L). The content of soluble carbohydrates (glucose, fructose, and sucrose) was found out on the basis of saccharide metabolism. Results showed an increase in glucose and sucrose concentrations in mature leaves treated with 100 μmol/L of Cd. Treatment with 20 μmol/L of Cd exhibited an increase in the concentration of sucrose and a decrease in the concentration of glucose and fructose. The mechanism of Cd interference with carbohydrate metabolism, especially the inhibition of the activity of invertase caused by Cd, was proposed as an explanation of soluble saccharides accumulation (Hédiji et al. 2010).

Rahoui et al. (2015) studied the effect of Cd on seedlings of six *Medicago truncatula* lines with different Cd susceptibility. The content of total soluble sugars, glucose, fructose, and sucrose was determined during 6 days of Cd treatment. Results showed that saccharide metabolism was a key component of Cd stress response. While Cd-tolerant lines of *M. truncatula* were characterized by high concentrations of glucose and/or sucrose in embryonic axes, high concentrations of fructose were determined in embryonic axes of susceptible lines of *M. truncatula*. The content of saccharides in susceptible lines was higher than in the tolerant lines. In contrast, tolerant lines showed higher mobilization of total soluble sugars and overconsumption of glucose under the toxic effect of Cd. Both lines affected sugar transport (Rahoui et al. 2015).

An increase in other saccharides (galactose, myoinositol, trehalose, and raffinose) was observed after Cd-treatment of *Arabidopsis thaliana* in the study of Sun et al. (2010).

Earlier studies suggested some pathways of Cd influence on plant saccharide metabolism: metallo-inhibition of saccharide transport in the bean and rice (Moya et al. 1993), Cd-caused reduction of saccharide transport and altering of α-amylase activity in cotyledon of the pea and the faba bean (Mihoub et al. 2005; Rahoui et al. 2008, 2010). Probably many other Cd-influenced mechanisms are taking place in plant organisms (Rahoui et al. 2015).

Since not only Cd has toxic effect on plant metabolism, studies using other heavy metals or multiple heavy metals together as noxious substances were performed.

The effect of Cd and Pb on growth and biochemical parameters was determined by John et al. (2008) on the water plant of *Lemna polyrhiza* L., which is known for its great ability to accumulate heavy metals. Plants were observed for their content of soluble sugars during 30 days of treatment with different concentrations of Cd and Pb. The findings showed that lower concentrations of Cd and Pb increased the content of soluble sugars. In contrast, higher concentrations of the metals (Cd >5 mg/kg of soil) decreased sugar content, which was probably caused by photosynthetic inhibition, by over-stimulation of respiration or by possible interaction with the active site of ribulosebisphosphate carboxylase (John et al. 2008).

The effect of Cu on the content of saccharides in 20-day-old cucumber plants was examined by Alaoui-Sossé et al. (2004). The content of starch and sucrose was measured in the first and second leaves and in roots. Starch was increased in both types of leaves when compared to controls, but the content of starch in the first leaves was significantly higher than in the second leaves. Starch was not detected in roots. The content of sucrose was similarly increased in both types of leaves, but it was not affected in roots. Saccharide accumulation may result from Cu inhibition of photosynthesis. However, it may also be a result of a decrease in phloem loading, a problem with the capacity of assimilate transport or it may be caused by a reduction of the utilization of nutrients including saccharides (Alaoui-Sossé et al. 2004).

Saccharides belong to significant nutrients in plant metabolism. They are the first products of CO_2 fixation, a precursor of the synthesis of many important organic compounds, they create energy for respiration, protect plants from wounds, infections, and stress situations, and they take care about detoxification pathways. Therefore, the disruption of saccharide metabolism may lead to a loss of the protective ability and structural integrity of the whole plant (Solanki and Dhankhar 2011).

8 Determination of Heavy Metals and Oxidative Stress

8.1 Determination of Heavy Metals in Biological Samples

The contamination of ecosystems and exposure to toxic metals is a significant worldwide burden. Therefore, biomonitoring techniques are getting more relevant because they may help to recognize contaminated area or crop, distribution of metals in the ecosystem and control of potential environmental hazards caused by heavy metal pollution (Ugulu 2015).

Major analytical techniques used to determine heavy metals in environmental matrices are Atomic Absorption Spectrometry (AAS), Inductively Coupled Plasma Atomic Emission Spectrometry (ICP/AES), Inductively Coupled Plasma Mass Spectrometry (ICP/MS), Neutron Activation Analysis (NAA), X-ray fluorescence (XRF), and Ion Chromatography (IC) (Li et al. 2015a, b; Markiewicz et al. 2015; Muller et al. 2015).

A large number of studies implemented so far to evaluate heavy-metal toxicity have focused on the development of analytical methods for the assessment of toxic species in diverse samples, mostly using techniques for elemental analysis, e.g. atomic absorption and emission spectroscopy, and inductively coupled plasma-mass spectrometry being probably the most widely used analytical technique for both multi-elemental analysis and speciation (Luque-Garcia et al. 2011).

In recent years, improvements in protein separation and identification methods and the progress of genomic knowledge have led to an enhancement in the utilization of proteomic techniques to answer biological questions (Isaacson et al. 2006). Plenty of investigations have shown that proteomics, in conjunction with bioinformatics tools, can facilitate the discovery of new, better biomarkers of metal exposure (Zhai et al. 2005).

Various proteomic approaches are accessible, including gel-based and gel-free methods. The most commonly used method in metal-toxicity-related proteomic studies is classical two-dimensional gel electrophoresis (2-DE), which has been predominantly coupled with Edman sequencing or peptide-mass fingerprinting (PMF) by means of a matrix-assisted laser desorption ionization-time-of-flight (MALDI-TOF) mass spectrometer (Luque-Garcia et al. 2011); a less frequent option is to use tandem MS (MS^2) {MALDI-TOF/TOF or LC-MS^2}, allowing sequencing of peptides and providing more confident protein identification.

A non-gel-based method, e.g. multidimensional protein identification technology (MudPIT), which involves the generation of peptides from a complex protein mixture, followed by two-dimensional chromatographic separation (Visioli et al. 2010), can partly resolve problems that are connected with the use of gel-based proteomic strategies.

According to Luque-Garcia et al. (2011), in the past decade, several quantitative proteomic strategies based on labeling proteins and peptides have been developed, e.g. isotope-coded affinity tag (ICAT), stable-isotope labeling by amino acids in cell culture (SILAC), isobaric tags for relative and absolute quantitation (iTRAQ), isotope-coded protein label (ICPL), N-terminal labeling, labeling with heavy water and even label-free techniques.

In general, the methods of analysis for the determination of heavy metals are changing to meet new or unforeseen conditions, and to ensure precision of measurement. The current analytical methods are precise enough to cover demand for the samples processing. On the other hand, prices of MS devices are still too high to be available for small labs or small companies, and research as well as introduction of simple measuring protocols is still desired. Last but not least important thing in actual literature is the limited number of analyses. As mentioned, modern/actual instrumentation and measurements are very expensive so the precious data for individual experiments are still very limited in number. Actual statistical and scientifically used programs like Statistica or C, C++, R-languages show perfectly values and trends of measured data but need still at least 7 analyses. However, X0 analyses are highly needed for appropriate modeling and better understanding of the real problems. For more details about statistical approach and modern uses of statistics in science, see, for example: Rencher (2002); Ostrouchov et al. (2012); Cass (2016).

8.2 Determination of Oxidative Stress

Three ways can be chosen for the determination of oxidative homeostasis and to judge whether homeostasis was disturbed and oxidative stress arose. Because ROS are not stable, they are not suitable as readily determined markers and therefore stable adducts. Therefore, products of radical mediated reactions and oxidized macromolecules serve as direct markers. Protein carbonyls, malondialdehyde as a product of radical degradation of lipids and 8-oxo-deoxyguanosine can be introduced as typical markers of oxidative damage (Yang et al. 2012; Pohanka 2013; Samsel et al. 2013). Malondialdehyde and protein carbonyls can be simply assayed in biological samples by spectrophotometry (Pohanka 2014c). 8-Oxo-deoxyguanosine can be determined by a competitive immunoassay like Enzyme-Linked Immunosorbent Assay (Gedik et al. 2002). Chromatography techniques for analysis of stress markers are available as well (Al-Rimawi 2015).

The occurrence of oxidative imbalance is followed by the expression of enzymes serving as antioxidants, i.e. enzymes with the ability to detoxify ROS. Such enzymes can be used as markers, and the emergence of oxidative stress can be deduced from their huge presence in plant tissues. Superoxide dismutase is a typical enzyme with significant antioxidant potency, and it is easily used as a marker of oxidative stress in plants (Cui et al. 2015; Jain et al. 2015; Rady and Hemida 2015). Catalase and peroxidase are other enzymes expressed under stress conditions (Naz et al. 2015).

Important information about a plant's condition and its ability to suppress oxidative stress can be learned from the measurement of the total level of low molecular weight antioxidants. In principle, the total level of antioxidants can be measured in two ways. Firstly, antioxidants can be identified as chemical entities and their amount is typically determined by chromatography (Abdennacer et al. 2015; Wang et al. 2015). Secondly, the total antioxidant capacity is measured by a simple technique. Assays named after the used reagent or principle of assay, such as ABTS [from 2,2′-azino-bis(3-ethylbenzothiazoline-6-sulfonic acid)], FRAP (from ferric reducing antioxidant power), ORAC (Oxygen Radical Absorbance Capacity), and DPPH (2,2-diphenyl-1-picrylhydrazyl) can be mentioned as examples (Ramirez-Anaya Jdel et al. 2015; Zhang et al. 2015). These approaches are simpler when compared to chromatography, and they achieve complex results. On the other hand, the information as to whether all or an isolated type of antioxidants are

Table 2 Markers of oxidative stress

Pathway	Exampled markers	References
Markers of oxidative damage	8-Oxo-deoxyguanosine, protein carbonyl, malondialdehyde	(Yang et al. 2012; Pohanka 2013; Samsel et al. 2013)
Enzymes expressed because of oxidative stress	Superoxide dismutase, catalase, peroxidase	(Cui et al. 2015; Jain et al. 2015; Rady and Hemida 2015)
Keeping of total antioxidant power	Low molecular weight antioxidants such as ascorbic acid, epigallocatechin gallate, quercetin	(El-Hawary et al. 2011)

depleted remains hidden. Ascorbic acid, epigallocatechin gallate, quercetin, and its derivatives are examples of typical plants' low molecular weight antioxidants (El-Hawary et al. 2011). Markers of oxidative stress are briefly summarized in Table 2.

9 Conclusions

Heavy metals represent a significant problem for the environment especially in former industrial areas. Long- term persistence of heavy metals is another problem which significantly worsens impact of heavy metals on the environment. Although most attention pertaining to this issue has been focused on their toxicity with respect to humans and animals, plants are not saved from the harmful impact of heavy metals either. Still, plants are potent in protecting themselves from heavy metals because of their ability to cumulate and distribute the metals into their body and store them for a long time. Undoubtedly, differences between certain plants and reasons why some plant species are more sensitive to the presence of metals or better at accumulating them have not been fully understood. The impact on oxidative homeostasis and the development of ROS during the exposure is probably one of significant effects mediated by heavy metals. However, more work on the issue should be done prior to making any definitive conclusions. Further research should be focused on both studying of metals impact on the plants including identification of pathological mechanisms, and establishing of processes where plants can serve as a tool for the metals remove from the environments.

10 Summary

Though toxicity of heavy metals is known in a junction to human health and animal laboratory models, plants appears to be out of the knowledge. It is a little surprising because environmental impact of heavy metals is also mediated through plants. In this paper, basic facts about heavy metals, their distribution in soil, mobility, accumulation by plants, and initiation of oxidative stress including the decline in basal metabolism are presented. The both actual and frontier studies in the field are summarized and discussed. The major pathophysiological pathways are introduced and known relations between heavy metals and their ability to initiate an oxidative damage are outlined for plants. Mobility and bioaccessibility are other factors that should be taken into consideration when heavy metals toxicity is evaluated and the both factors are discussed here. This review contains a wide discussion about metals like lead, mercury, copper, cadmium, iron, zinc, nickel, and vanadium. This survey can be concluded by a statement that heavy metals are significant contributors to pathological processes in most of the known and studied plants and oxidative stress takes place in these processes.

Acknowledgments This work was supported by the Ministry of Education, Youth and Sports of the Czech Republic under the CEITEC 2020 (LQ1601) and NAZV QJ1320122 projects. We thank all reviewers for their constructive reviews of this manuscript, and Prof. Pim de Voogt for editorial comments that led to improvements of the manuscript.

References

Abdennacer B, Karim M, Yassine M, Nesrine R, Mouna D, Mohamed B (2015) Determination of phytochemicals and antioxidant activity of methanol extracts obtained from the fruit and leaves of Tunisian *Lycium intricatum* Boiss. Food Chem 174:577–584. https://doi.org/10.1016/j.foodchem.2014.11.114

Adrees M, Ali S, Iqbal M et al (2015) Mannitol alleviates chromium toxicity in wheat plants in relation to growth, yield, stimulation of anti-oxidative enzymes, oxidative stress and Cr uptake in sand and soil media. Ecotox Environ Safe 122:1–8. https://doi.org/10.1016/j.ecoenv.2015.07.003

Agrawal B, Czymmek KJ, Sparks DL, Bais HP (2013) Transient influx of nickel in root mitochondria modulates organic acid and reactive oxygen species production in nickel hyperaccumulator *Alyssum murale*. J Biol Chem 288(10):7351–7362. https://doi.org/10.1074/jbc.M112.406645

Ahmad R, Tehsin Z, Malik ST et al (2016) Phytoremediation potential of hemp (*Cannabis sativa* L.): identification and characterization of heavy metals responsive genes. Clean-Soil Air Water 44(2):195–201. https://doi.org/10.1002/clen.201500117

Alaoui-Sossé B, Genet P, Vinit-Dunand F, Toussaint ML, Epron D, Badot PM (2004) Effect of copper on growth in cucumber plants (*Cucumis sativus*) and its relationships with carbohydrate accumulation and changes in ion contents. Plant Sci 166:1213–1218. https://doi.org/10.1016/j.plantsci.2003.12.032

Al-Rimawi F (2015) Development and validation of a simple reversed-phase HPLC-UV method for determination of malondialdehyde in olive oil. J Am Oil Chem Soc 92(7):933–937. https://doi.org/10.1007/s11746-015-2664-x

Amor Y, Haigle CH, Johnson S, Wainscott M, Delmer DP (1995) A membrane-associated form of sucrose synthase and its potential role in synthesis of cellulose and callose in plants. Proc Natl Acad Sci U S A 92(20):9353–9357. https://doi.org/10.1073/pnas.92.20.9353

Anjum NA, Hasanuzzaman M, Hossain MA et al (2015) Jacks of metal/metalloid chelation trade in plants-an overview. Front Plant Sci 6(192):1–17. https://doi.org/10.3389/fpls.2015.00192

Antosiewicz DM, Barabasz A, Siemianowski O (2014) Phenotypic and molecular consequences of overexpression of metal-homeostasis genes. Front Plant Sci 5(80):1–7. https://doi.org/10.3389/fpls.2014.00080

Assunção AGL, Schat H, Aarts MGM (2003) *Thlaspi caerulescens*, an attractive model species to study heavy metal hyperaccumulation in plants. New Phytol 159(2):351–360. https://doi.org/10.1046/j.1469-8137.2003.00820.x

Barman SC, Sahu RK, Bhargava SK, Chaterjee C (2000) Distribution of heavy metals in wheat, mustard, and weed grown in field irrigated with industrial effluents. Bull Environ Contam Toxicol 64(4):489–496. https://doi.org/10.1007/s001280000030

Bhargava A, Carmona FF, Bhargava M, Srivastava S (2012) Approaches for enhanced phytoextraction of heavy metals. J Environ Manage 105:103–120. https://doi.org/10.1016/j.jenvman.2012.04.002

Brown RM Jr, Saxena IM, Kudlicka K (1996) Cellulose biosynthesis in higher plants. Trends Plant Sci 1:149–156. doi: https://doi.org/10.1016/S1360-1385(96)80050-1

Brunner I, Luster J, Günthardt-Goerg MS, Frey B (2008) Heavy metal accumulation and phytostabilisation potential of tree fine roots in a contaminated soil. Environ Pollut 152 (3):559–568. https://doi.org/10.1016/j.envpol.2007.07.006

Cass S (2016) The 2016 top programming languages. http://spectrum.ieee.org/computing/software/the-2016-top-programming-languages

Chettri MK, Cook CM, Vardaka E, Sawidis T, Lanaras T (1998) The effect of Cu, Zn and Pb on the chlorophyll content of the lichens *Cladonia convoluta* and *Cladonia rangiformis*. Environ Exp Bot 39(1):1–10. https://doi.org/10.1016/S0098-8472(97)00024-5

Clemens S (2006) Toxic metal accumulation, responses to exposure and mechanisms of tolerance in plants. Biochimie 88(11):1707–1719. https://doi.org/10.1016/j.biochi.2006.07.003

Cobbet C, Goldsbrough P (2002) Phytochelatins and metallothioneins: roles in heavy metal detoxification and homeostasis. Annu Rev Plant Physiol 53:159–182. https://doi.org/10.1146/annurev.arplant.53.100301.135154

Colzi I, Doumett S, Del Bubba M et al (2011) On the role of the cell wall in the phenomenon of copper tolerance in *Silene paradoxa* L. Environ Exp Bot 72:77–83. https://doi.org/10.1016/j.envexpbot.2010.02.006

Cui LJ, Huang Q, Yan B et al (2015) Molecular cloning and expression analysis of a Cu/Zn SOD gene (BcCSD1) from *Brassica campestris ssp chinensis*. Food Chem 186:306–311. https://doi.org/10.1016/j.foodchem.2014.07.121

Cumino A, Curatti L, Giarrocco L, Salerno GL (2002) Sucrose metabolism: anabaena sucrose-phosphate synthase and sucrose-phosphate phosphatase define minimal functional domains shuffled during evolution. FEBS Lett 517(1–3):19–23. https://doi.org/10.1016/S0014-5793(02)02516-4

Das P, Samantaray S, Rout GR (1997) Studies on cadmium toxicity in plants: a review. Environ Pollut 98(1):29–36. https://doi.org/10.1016/S0269-7491(97)00110-3

De Abreu-Neto JB, Turchetto-Zolet AC, De Oliveira LF, Zanettini MH, Margis-Pinheiro M (2013) Heavy metal-associated isoprenylated plant protein (HIPP): characterization of a family of proteins exclusive to plants. FEBS J 280(7):1604–1616. https://doi.org/10.1111/febs.12159

De Matos AT, Fontes MPF, Da Costa LM, Martinez MA (2001) Mobility of heavy metals as related to soil chemical and mineralogical characteristics of Brazilian soils. Environ Pollut 111 (3):429–435. https://doi.org/10.1016/S0269-7491(00)00088-9

Debbaudt AL, Ferreira ML, Gschaider ME (2004) Theoretical and experimental study of M^{2+} adsorption on biopolymers. III. Comparative kinetic pattern of Pb, Hg and Cd. Carbohydr Polym 56(3):321–332. https://doi.org/10.1016/j.carbpol.2004.02.009

Demidchik V (2015) Mechanisms of oxidative stress in plants: from classical chemistry to cell biology. Environ Exp Bot 109:212–228. https://doi.org/10.1016/j.envexpbot.2014.06.021

Dimkpa CHO, Merten D, Svatos A, Büchel G, Kothe E (2009) Metal-induced oxidative stress impacting plant growth in contaminated soil is alleviated by microbial siderophores. Soil Biol Biochem 41(1):154–162. https://doi.org/10.1016/j.soilbio.2008.10.010

Duffus JH (2002) "Heavy metals" a meaningless term? IUPAC Tech Rep Pure Appl Chem 74 (5):793–807. https://doi.org/10.1351/pac200274050793

Duffus JH (2003) Errata "heavy metals" a meaningless term? IUPAC Tech Rep Pure Appl Chem 75(9):1357. https://doi.org/10.1351/pac200375091357

El-Hawary SA, Sokkar NM, Ali ZY, Yehia MM (2011) A profile of bioactive compounds of *Rumex vesicarius* L. J Food Sci 76(8):1195–1202. https://doi.org/10.1111/j.1750-3841.2011.02370.x

Emamverdian A, Ding Y, Mokhberdoran F, Xie Y (2015) Heavy metal stress and some mechanisms of plant defense response. Sci World J 2015:1–18. https://doi.org/10.1155/2015/756120

Ettler V, Kribek B, Majer V, Knesl I, Mihaljevic M (2012) Differences in the bioaccessibility of metals/metalloids in soils from mining and smelting areas (Copperbelt, Zambia). J Geochem Explor 113:68–75. https://doi.org/10.1016/j.gexplo.2011.08.001

Farahat E, Linderholm HW (2015) The effect of long-term wastewater irrigation on accumulation and transfer of heavy metals in *Cupressus sempervirens* leaves and adjacent soils. Sci Total Environ 512–513:1–7. https://doi.org/10.1016/j.scitotenv.2015.01.032

Fargasova A (2004) Toxicity comparison of some possible toxic metals (Cd, Cu, Pb, Se, Zn) on young seedlings of *Sinapis alba* L. Plant Soil Environ 50(1):33–38

Fernandez LR, Vandenbussche G, Roosens N, Govaerts C, Goormaghtigh E, Verbruggen N (2012) Metal binding properties and structure of a type III metallothionein from the metal hyperaccumulator plant *Noccaea caerulescens*. Biochim Biophys Acta 1824(9):1016–1023. https://doi.org/10.1016/j.bbapap.2012.05.010

Flouty R, Khalaf G (2015) Role of Cu and Pb on Ni bioaccumulation by Chlamydomonas reinhardtii: validation of the biotic ligand model in binary metal mixtures. Ecotox Environ Safe 113:79–86. https://doi.org/10.1016/j.ecoenv.2014.11.022

Franco CR, Chagas AP, Jorge RA (2004) Ion-exchange equilibria with aluminum pectinates. Colloid Surf A 204(1–3):183–192. https://doi.org/10.1016/S0927-7757(01)01134-7

Freeman JL, Garcia D, Kim D, Hopf A, Salt DE (2005) Constitutively elevated salicylic acid signals glutathione-mediated nickel tolerance in *Thlaspi* nickel hyperaccumulators. Plant Physiol 137(3):1082–1091. https://doi.org/10.1104/pp.104.055293

Freeman JL, Tamaoki M, Stushnoff C et al (2010) Molecular mechanisms of selenium tolerance and hyperaccumulation in *Stanleya pinnata*. Plant Physiol 153(4):1630–1652. https://doi.org/10.1104/pp.110.156570

Gajewska E, Głowacki R, Mazur J, Skłodowska M (2013) Differential response of wheat roots to Cu, Ni and Cd treatment: oxidative stress and defense reactions. Plant Growth Regul 71 (1):13–20. https://doi.org/10.1007/s10725-013-9803-x

Galfati I, Bilal E, Sassi AB, Abdallah H, Zaier A (2011) Accumulation of heavy metals in native plants growing near the phosphate treatment industry, Tunisia. Carpath J Earth Env 6 (2):85–100

Garcia L, Welchen E, Gonzalez DH (2014) Mitochondria and copper homeostasis in plants. Mitochondrion 19:269–274. https://doi.org/10.1016/j.mito.2014.02.011

Gedik CM, Boyle SP, Wood SG, Vaughan NJ, Collins AR (2002) Oxidative stress in humans: validation of biomarkers of DNA damage. Carcinogenesis 23(9):1441–1446. https://doi.org/10.1093/carcin/23.9.1441

Gendre D, Czernic P, Conéjéro G et al (2006) TcYSL3, a member of the YSL gene family from the hyperaccumulator *Thlaspi caerulescens*, encodes a nicotianamine-Ni/Fe transporter. Plant J 49 (1):1–15. https://doi.org/10.1111/j.1365-313X.2006.02937.x

Guo WJ, Meetam M, Goldsbrough PB (2008) Examining the specific contributions of individual *Arabidopsis* metallothioneins to copper distribution and metal tolerance. Plant Physiol 146:1697–1706. https://doi.org/10.1104/pp.108.115782

Guo X, Meng H, Zhu S, Zhang T, Yu S (2015) Purifying sugar beet pectins from non-pectic components by means of metal precipitation. Food Hydrocoll 51:69–75. https://doi.org/10.1016/j.foodhyd.2015.05.009

Gupta AK, Ahmad M (2013) Effect of refinery waste effluent on tocopherol, carotenoid, phenolics and other antioxidants content in Allium cepa. Toxicol Ind Health 29(7):652–661. https://doi.org/10.1177/0748233712436639

He F, Liu Q, Zheng L, Cui Y, Shen Z, Zheng L (2015) RNA-Seq analysis of rice roots reveals the involvement of post-transcriptional regulation in response to cadmium stress. Front Plant Sci 6 (1136):1–16. https://doi.org/10.3389/fpls.2015.01136

Hechmi N, Ben Aissa N, Abdenaceur H, Jedidi N (2015) Uptake and bioaccumulation of pentachlorophenol by emergent wetland plant Phragmites australis (common reed) in cadmium co-contaminated soil. Int J Phytoremediation 17:109–116. https://doi.org/10.1080/15226514.2013.851169

Hédiji H, Djebali W, Cabasson C et al (2010) Effects of long-term cadmium exposure on growth and metabolomic profile of tomato plants. Ecotox Environ Safe 73(8):1965–1974. https://doi.org/10.1016/j.ecoenv.2010.08.014

Hieu HC, Li H, Miyauchi Y, Mizutani G, Fujita N, Nakamura Y (2015) Wetting effect on optical sum frequency generation (SFG) spectra of D-glucose, D-fructose, and sucrose. Spectrochim Acta A Mol Biomol Spectrosc 138:834–839. https://doi.org/10.1016/j.saa.2014.10.108

Hossain Z, Komatsu S (2013) Contribution of proteomic studies towards understanding plant heavy metal stress response. Front Plant Sci 3(310):1–12. https://doi.org/10.3389/fpls.2012.00310

Hu Y, Wang T, Yang X, Zhao Y (2014) Analysis of compositional monosaccharides in fungus polysaccharidesby capillary zone electrophoresis. Carbohydr Polym 102:481–488. https://doi.org/10.1016/j.carbpol.2013.11.054

Isaacson T, Damasceno CMB, Saravanan RS et al (2006) Sample extraction techniques for enhanced proteomic analysis of plant tissues. Nat Protoc 1(2):769–774. https://doi.org/10.1038/nprot.2006.102

Jain R, Chandra A, Venugopalan VK, Solomon S (2015) Physiological changes and expression of SOD and P5CS genes in response to water deficit in sugarcane. Sugar Tech 17(3):276–282. https://doi.org/10.1007/s12355-014-0317-2

John R, Ahmad P, Gadgil K, Sharma S (2008) Effect of cadmium and lead on growth, biochemical parameters and uptake in *Lemna polyrrhiza* L. Plant Soil Environ 54(6):262–270

Jomova K, Valko M (2011) Advances in metal-induced oxidative stress and human disease. Toxicology 283(2–3):65–87. https://doi.org/10.1016/j.tox.2011.03.001

Juknys R, Vitkauskaitė G, Račaitė M, Venclovienė J (2012) The impacts of heavy metals on oxidative stress and growth of spring barley. Cent Eur J Biol 7(2):299–306. https://doi.org/10.2478/s11535-012-0012-9

Kalubi KN, Mehes-Smith M, Omri A (2016) Comparative analysis of metal translocation in red maple (*Acer rubrum*) and trembling aspen (*Populus tremuloides*) populations from stressed ecosystems contaminated with metals. Chem Ecol 32(4):312–323. https://doi.org/10.1080/02757540.2016.1142978

Kartel MT, Kupchik LA, Veisov BK (1999) Evaluation of pectin binding of heavy metal ions in aqueous solutions. Chemosphere 38(11):2591–2596. https://doi.org/10.1016/S0045-6535(98)00466-4

Kim RY, Yoon JK, Kim TS, Yang JE, Owns G, Kim KR (2015) Bioavailability of heavy metals in soils: definitions and practical implementation—a critical review. Environ Geochem Health 37 (6):1041–1061. https://doi.org/10.1007/s10653-015-9695-y

Kirkham MB (2006) Cadmium in plants on polluted soils: effects of soil factors, hyperaccumulation, and amendments. Geoderma 137(1):19–32. https://doi.org/10.1016/j.geoderma.2006.08.024

Knasmüller S, Gottmann E, Steinkellner H et al (1998) Detection of genotoxic effects of heavy metal contaminated soils with plant bioassays. Mutat Res 420(1–3):37–48. https://doi.org/10.1016/S1383-5718(98)00145-4

Koch K (2004) Sucrose metabolism: regulatory mechanisms and pivotal roles in sugar sensing and plant development. Curr Opin Plant Biol 7(3):235–246. https://doi.org/10.1016/j.pbi.2004.03.014

Krämer U, Talke IN, Hanikenne M (2007) Transition metal transport. FEBS Lett 581 (12):2263–2272. https://doi.org/10.1016/j.febslet.2007.04.010

Kudlicka K, Brown RM Jr (1997) Cellulose and Callose biosynthesis in higher plants. Plant Physiol 115(2):643–656. https://doi.org/10.1104/pp.115.2.643

Kumar N, Bauddh K, Kumar S, Dwivedi N, Singh DP, Barman SC (2013) Accumulation of metals in weed species grown on the soil contaminated with industrial waste and their phytoremediation potential. Ecol Eng 61:491–495. https://doi.org/10.1016/j.ecoleng.2013.10.004

Kurd F, Samavati V (2015) Water soluble polysaccharides from *Spirulina platensis*: extraction and in vitro anti-cancer activity. Int J Biol Macromol 74:498–506. https://doi.org/10.1016/j.ijbiomac.2015.01.005

Li E, AC W, Li J, Liu Q, Gilbert RG (2015a) Improved understanding of rice amylose biosynthesis from advanced starch structural characterization. Rice 8(20):1–8. https://doi.org/10.1186/s12284-015-0055-4

Li ZM, Yu Y, Li ZL (2015b) A review of biosensing techniques for detection of trace carcinogen contamination in food products. Anal Bioanal Chem 407(10):2711–2726. https://doi.org/10.1007/s00216-015-8530-8

Liang HM, Lin TH, Chiou JM, Yeh KC (2009) Model evaluation of the phytoextraction potential of heavy metal hyperaccumulators and non-hyperaccumulators. Environ Pollut 157 (6):1945–1952. https://doi.org/10.1016/j.envpol.2008.11.052

Lin YF, Severing EI, Te Lintel Hekkert B, Schijlen E, Aarts MGM (2014) A comprehensive set of transcript sequences of the heavy metal hyperaccumulator Noccaea caerulescens. Front Plant Sci 5(261):1–15. https://doi.org/10.3389/fpls.2014.00261

Lou Y, Yang Y, Hu L, Liu H, Xu Q (2015) Exogenous glycinebetaine alleviates the detrimental effect of Cd stress on perennial ryegrass. Ecotoxicology 24(6):1330–1340. https://doi.org/10.1007/s10646-015-1508-7

Luque-Garcia JL, Cabezas-Sanchez P, Camara C (2011) Proteomics as a tool for examining the toxicity of heavy metals. Trends Anal Chem 30(5):703–716. https://doi.org/10.1016/j.trac.2011.01.014

Lv Y, Deng X, Quan L, Xia Y, Shen Z (2013) Metallothioneins BcMT1 and BcMT2 from *Brassica campestris* enhance tolerance to cadmium and copper and decrease production of reactive oxygen species in *Arabidopsis thaliana*. Plant and Soil 367(1–2):507–519. https://doi.org/10.1007/s11104-012-1486-y

Ma JF (2005) Plant root responses to three abundant soil minerals: silicon, aluminum and iron. Crit Rev Plant Sci 24(4):267–281. https://doi.org/10.1080/07352680500196017

Markiewicz B, Komorowicz I, Belter M, Baralkiewicz D (2015) Chromium and its speciation in water samples by HPLC/ICP-MS—technique establishing metrological traceability: a review since 2000. Talanta 132:814–828. https://doi.org/10.1016/j.talanta.2014.10.002

Mata YN, Blázquez ML, Ballester A, González F, Muñoz JA (2009) Sugar-beet pulp pectin gels as biosorbent for heavy metals: preparation and determination of biosorption and desorption characteristics. Chem Eng J 150(2–3):289–301. https://doi.org/10.1016/j.cej.2009.01.001

Mehes-Smith M, Nkongolo KK, Narendrula R, Cholewa E (2013a) Mobility of heavy metals in plants and soil: a case study from a mining region in Canada. Am J Environ Sci 9(6):483–493. https://doi.org/10.3844/ajessp.2013.483.493

Mehes-Smith M, Nkongolo KK, Cholewa E (2013b) Coping mechanisms of plants to metal contaminated soil. In: Silvern S, Young S (eds) Environmental change and sustainability. Chapter 3, pp 53–90. InTech. ISBN: 978-953-51-1094-1. doi: https://doi.org/10.5772/55124

Mejáre M, Bülow L (2001) Metal binding proteins and peptides in bioremediation and phytoremediation of heavy metals. Trends Biotechnol 19(2):67–73. https://doi.org/10.1016/S0167-7799(00)01534-1

Mera R, Torres E, Abalde J (2016) Influence of sulphate on the reduction of cadmium toxicity in the microalga Chlamydomonas moewusii. Ecotox Environ Safe 128:236–245. https://doi.org/10.1016/j.ecoenv.2016.02.030

Mesjasz-Przybyłowicz J, Barnabas A, Przybyłowicz W (2007) Comparison of cytology and distribution of nickel in roots of Ni-hyperaccumulating and non-hyperaccumulating genotypes of *Senecio coronatus*. Plant and Soil 293:61–78. https://doi.org/10.1007/s11104-007-9237-1

Mganga N, Manoko MLK, Rulangaranga ZK (2011) Classification of plants according to their heavy metal content around North Mara gold mine, Tanzania: implication for phytoremediation. Tanz J Sci 37:109–119

Mihoub A, Chaoui A, El Ferjani E (2005) Biochemical changes associated with cadmium and copper stress in germinating pea seeds (*Pisum sativum* L.) C R Biol 328(1):33–41. https://doi.org/10.1016/j.crvi.2004.10.003

Miransari M (2011) Hyperaccumulators, arbuscular mycorrhizal fungi and stress of heavy metals. Biotechnol Adv 29(6):645–653. https://doi.org/10.1016/j.biotechadv.2011.04.006

Moya JL, Ros R, Picazo I (1993) Influence of cadmium and nickel on growth, photosynthesis and carbohydrate distribution in rice plants. Photosynth Res 36(2):75–80. https://doi.org/10.1007/BF00016271

Muller AL, Oliveira JS, Mello PA, Muller EI, Flores EM (2015) Study and determination of elemental impurities by ICP-MS in active pharmaceutical ingredients using single reaction chamber digestion in compliance with USP requirements. Talanta 136:161–169. https://doi.org/10.1016/j.talanta.2014.12.023

Mussatto SI, Mancilha IM (2007) Non-digestible oligosaccharides: a review. Carbohydr Polym 68 (3):587–597. https://doi.org/10.1016/j.carbpol.2006.12.011

Myers AM, Morell MK, James MG, Ball SG (2000) Recent progress toward understanding biosynthesis of the amylopectin crystal. Plant Physiol 122:989–997. https://doi.org/10.1104/pp.122.4.989

Nahar K, Hasanuzzaman M, Alam MM, Fujita M (2015) Exogenous glutathione confers high temperature stress tolerance in mung bean (*Vigna radiata* L.) by modulating antioxidant defense and methylglyoxal detoxification system. Environ Exp Bot 112:44–54. https://doi.org/10.1016/j.envexpbot.2014.12.001

Naz FS, Yusuf M, Khan TA, Fariduddin Q, Ahmad A (2015) Low level of selenium increases the efficacy of 24-epibrassinolide through altered physiological and biochemical traits of *Brassica juncea* plants. Food Chem 185:441–448. https://doi.org/10.1016/j.foodchem.2015.04.016

Ostrouchov G, Chen W-C, Schmidt D, Patel P (2012) Programming with big data in R. http://r-pbd.org/

Park W, Feng Y, Ahn SJ (2014) Alteration of leaf shape, improved metal tolerance, and productivity of seed by overexpression of CsHMA3 in *Camelina sativa*. Biotechnol Biofuels 7 (96):1–17. https://doi.org/10.1186/1754-6834-7-96

Perrin RM (2001) Cellulose: how many cellulose synthases to make a plant? Curr Biol 11 (6):213–216. https://doi.org/10.1016/S0960-9822(01)00108-7

Pohanka M (2013) Role of oxidative stress in infectious diseases. A review. Folia Microbiol 58 (6):503–513. https://doi.org/10.1007/s12223-013-0239-5

Pohanka M (2014a) Alzheimer's disease and oxidative stress. A review. Curr Med Chem 21 (3):356–364. https://doi.org/10.2174/09298673113206660258

Pohanka M (2014b) Copper, aluminum, iron and calcium inhibit human acetylcholinesterase in vitro. Environ Toxicol Pharmacol 37(1):455–459. https://doi.org/10.1016/j.etap.2014.01.001

Pohanka M (2014c) Caffeine alters oxidative homeostasis in the body of BALB/c mice. Bratisl Med J 115(11):699–703. https://doi.org/10.4149/BLL_2014_135

Puls RW, Powell RM, Clark D, Eldred CJ (1991) Effects of pH, solid/solution ratio, ionic strength and organic acids on Pb and Cd sorption on kaolinite. Water Air Soil Pollut 57–58(1):423–430. https://doi.org/10.1007/BF00282905

Rady MM, Hemida KA (2015) Modulation of cadmium toxicity and enhancing cadmium-tolerance in wheat seedlings by exogenous application of polyamines. Ecotoxic Environ Safe 119:178–185. https://doi.org/10.1016/j.ecoenv.2015.05.008

Rahoui S, Chaoui A, El Ferjani E (2008) Differential sensitivity to cadmium in germinating seeds of three cultivars of faba bean (*Vicia faba* L.) Acta Physiol Plant 30(4):451–456. https://doi.org/10.1007/s11738-008-0142-x

Rahoui S, Chaoui A, El Ferjani E (2010) Reserve mobilization disorder in germinating seeds of *Vicia faba* exposed to cadmium. J Plant Nutr 33(5–8):809–817. https://doi.org/10.1080/01904161003654055

Rahoui S, Chaoui A, Ben C, Rickauer M, Gentzbittel L, El Ferjani E (2015) Effect of cadmium pollution on mobilization of embryo reserves in seedlings of six contrasted *Medicago truncatula* lines. Phytochemistry 111:98–106. https://doi.org/10.1016/j.phytochem.2014.12.002

Ramirez-Anaya Jdel P, Samaniego-Sanchez C, Castañeda-Saucedo MC, Villalon-Mir M, López-García de la Serrana H (2015) Phenols and the antioxidant capacity of Mediterranean vegetables prepared with extra virgin olive oil using different domestic cooking techniques. Food Chem 188:430–438. https://doi.org/10.1016/j.foodchem.2015.04.124

Rascio N, Navari-Izzo F (2011) Heavy metal hyperaccumulating plants: how and why do they do it? And what makes them so interesting? Plant Sci 180(2):169–181. https://doi.org/10.1016/j.plantsci.2010.08.016

Rees F, Germain C, Sterckeman T, Morel JL (2015) Plant growth and metal uptake by a non-hyperaccumulating species (*Lolium perenne*) and a Cd-Zn hyperaccumulator (*Noccaea caerulescens*) in contaminated soils amended with biochar. Plant Soil 395(1–2):57–73. https://doi.org/10.1007/s11104-015-2384-x

Reiter WD (2002) Biosynthesis and properties of the plant cell wall. Curr Opin Plant Biol 5 (6):536–542. https://doi.org/10.1016/S1369-5266(02)00306-0

Reiter WD, Vanzin GF (2001) Molecular genetics of nucleotide sugar interconversion pathways in plants. Plant Mol Biol 47(1–2):95–113. https://doi.org/10.1007/s11104-015-2384-x

Rencher AC (2002) Methods of multivariate analysis. Wiley series in probability and mathematical statistics. Wiley, Hoboken, NJ

Ricachenevsky FK, Menguer PK, Sperotto RA, Williams LE, Fett JP (2013) Roles of plant metal tolerance proteins (MTP) in metal storage and potential use in biofortification strategies. Front Plant Sci 4(144):1–16. https://doi.org/10.3389/fpls.2013.00144

Rorabaugh JM, Stratford JM, Zahniser NR (2015) Differences in bingeing behavior and cocaine reward following intermittent access to sucrose, glucose or fructose solutions. Neuroscience 301:213–220. https://doi.org/10.1016/j.neuroscience.2015.06.015

Sainger PA, Dhankhar R, Sainger M, Kaushik A, Singh RP (2011) Assessment of heavy metal tolerance in native plant species from soils contaminated with electroplating effluent. Ecotox Environ Safe 74(8):2284–2291. https://doi.org/10.1016/j.ecoenv.2011.07.028

Salerno GL, Curatti L (2003) Origin of sucrose metabolism in higher plants: when, how and why? Trends Plant Sci 8(2):63–69. https://doi.org/10.1016/S1360-1385(02)00029-8

Samsel M, Dzierzbicka K, Trzonkowski P (2013) Adenosine, its analogues and conjugates. Postepy Hig Med Dosw 67:1189–1203. https://doi.org/10.5604/17322693.1078588

Sarma H (2011) Metal hyperaccumulation in plants: a review focusing on phytoremediation technology. J Environ Sci Technol 4(2):118–138. https://doi.org/10.3923/jest.2011.118.138

Schützendübel A, Polle A (2001) Plant responses to abiotic stresses: heavy metal induced oxidative stress and protection by mycorrhization. J Exp Bot 53(372):1351–1365. https://doi.org/10.1093/jexbot/53.372.1351

Seth CS, Remans T, Keunen E et al (2012) Phytoextraction of toxic metals: a central role for glutathione. Plant Cell Environ 35(2):334–346. https://doi.org/10.1111/j.1365-3040.2011.02338.x

Sethy SK, Ghosh S (2013) Effect of heavy metals on germination of seeds. J Nat Sci Biol Med 4 (2):272–275. https://doi.org/10.4103/0976-9668.116964

Sharma SS, Dietz KJ (2009) The relationship between metal toxicity and cellular redox imbalance. Trends Plant Sci 14(1):43–50. https://doi.org/10.1016/j.tplants.2008.10.007

Sharma P, Jha AB, Dubey RS, Pessarakli M (2012) Reactive oxygen species, oxidative damage, and antioxidative defense mechanism in plants under stressful conditions. J Bot 2012:1–26. https://doi.org/10.1155/2012/217037

Siemianowski O, Barabasz A, Kendziorek M et al (2014) HMA4 expression in tobacco reduces Cd accumulation due to the induction of the apoplastic barrier. J Exp Bot 65(4):1125–1139. https://doi.org/10.1093/jxb/ert471

Singh OV, Labana S, Pandey G, Budhiraja R (2003) Phytoremediation: an overview of metallic ion decontamination from soil. Appl Microbiol Biotechnol 61(5–6):405–412. https://doi.org/10.1007/s00253-003-1244-4

Slewinski TL, Braun DM (2010) Current perspectives on the regulation of whole-plant carbohydrate partitioning. Plant Sci 178(4):341–349. https://doi.org/10.1016/j.plantsci.2010.01.010

Solanki R, Dhankhar R (2011) Biochemical changes and adaptive strategies of plants under heavy metal stress. Biologia 66(2):195–204. https://doi.org/10.2478/s11756-011-0005-6

Stankovic S, Kalaba P, Stankovic AR (2014) Biota as toxic metal indicators. Environ Chem Lett 12(1):63–84. https://doi.org/10.1007/s10311-013-0430-6

Sun X, Zhang J, Zhang H et al (2010) The responses of *Arabidopsis thaliana* to cadmium exposure explored via metabolite profiling. Chemosphere 78(7):840–845. https://doi.org/10.1016/j.chemosphere.2009.11.045

Tamoi M, Nagaoka M, Miyagawa Y, Shigeoka S (2006) Contribution of fructose-1,6-bisphosphatase and sedoheptulose-1,7-bisphosphatase to the photosynthetic rate and carbon flow in the Calvin cycle in transgenic plants. Plant Cell Physiol 47(3):380–390. https://doi.org/10.1093/pcp/pcj004

Topolska J, Latowski D, Kaschabek S, Manecki M, Merkel BJ, Rakovan J (2014) Pb remobilization by bacterially mediated dissolution of pyromorphite Pb5(PO4)3Cl in presence of phosphate-solubilizing *Pseudomonas putida*. Environ Sci Pollut R 21(2):1079–1089. https://doi.org/10.1007/s11356-013-1968-3

Ugulu I (2015) Determination of heavy metal accumulation in plant samples by spectrometric techniques in Turkey. Appl Spectrosc Rev 50(2):113–151. https://doi.org/10.1080/05704928.2014.935981

Van Bussel CGJ, Schroeder JP, Mahlmann L, Schulz C (2014) Aquatic accumulation of dietary metals (Fe, Zn, Cu, Co, Mn) in recirculating aquaculture systems (RAS) changes body composition but not performance and health of fuvenile turbot (*Psetta maxima*). Aquacult Eng 61:35–42. https://doi.org/10.1016/j.aquaeng.2014.05.003

Vatamaniuk OK, Bucher EA, Ward JT, Rea PA (2001) A new pathway for heavy metal detoxification in animals. Phytochelatin synthase is required for cadmium tolerance in Caenorhabditis elegans. J Biol Chem 276:20817–20820. https://doi.org/10.1074/jbc.C100152200

Velisek J, Cejpek K (2005a) Biosynthesis of food constituents: saccharides. 1. Monosaccharides, oligosaccharides, and related compounds—a review. Czech J Food Sci 23(4):129–144

Velisek J, Cejpek K (2005b) Biosynthesis of food constituents: saccharides. 2. Polysaccharides—a review. Czech J Food Sci 23(5):173–183

Verbruggen N, Hermans C, Schat H (2009) Molecular mechanisms of metal hyperaccumulation in plants. New Phytol 181(4):759–776. https://doi.org/10.1111/j.1469-8137.2008.02748.x

Verret F, Gravot A, Auroy P et al (2004) Overexpression of AtHMA4 enhances root-to-shoot translocation of zinc and cadmium and plant metal tolerance. FEBS Lett 576(3):306–312. https://doi.org/10.1016/j.febslet.2004.09.023

Violante A, Cozzolino V, Perelomov L, Caporale AG, Pigna M (2010) Mobility and bioavailability of heavy metals and metalloids in soil environments. J Soil Sci Plant Nutr 10(3):268–292. https://doi.org/10.4067/S0718-95162010000100005

Visioli G, Marmiroli M, Marmiroli N (2010) Two-dimensional liquid chromatography technique coupled with mass spectrometry analysis to compare the proteomic response to cadmium stress in plants. J Biomed Biotechnol 2010:1–10. https://doi.org/10.1155/2010/567510

Visioli G, D'Egidio S, Vamerali T, Mattarozzi M, Sanangelantoni AM (2014) Culturable endophytic bacteria enhance Ni translocation in the hyperaccumulator *Noccaea caerulescens*. Chemosphere 117:538–544. https://doi.org/10.1016/j.chemosphere.2014.09.014

Wang T, Sun H (2013) Biosorption of heavy metals from aqueous solution by UV-mutant *Bacillus subtilis*. Environ Sci Pollut R 20(10):7450–7463. https://doi.org/10.1007/s11356-013-1767-x

Wang C, Hua D, Yan C (2015) Structural characterization and antioxidant activities of a novel fructan from *Achyranthes bidentata* Blume, a famous medicinal plant in China. Ind Crop Prod 70:427–434. https://doi.org/10.1016/j.indcrop.2015.03.051

Wasserman LA, Sergeev AI, Vasil'ev VG et al (2015) Thermodynamic and structural properties of tuber starches from transgenic potato plants grown in vitro and in vivo. Carbohydr Polym 125:214–223. https://doi.org/10.1016/j.carbpol.2015.01.084

Wei JL, Lai HY, Chen ZS (2012) Chelator effects on bioconcentration and translocation of cadmium by hyperaccumulators, *Tagetes patula* and *Impatiens walleriana*. Ecotox Environ Safe 84:173–178. https://doi.org/10.1016/j.ecoenv.2012.07.004

Wingler A (2002) The function of trehalose biosynthesis in plants. Phytochemistry 60(5):437–440. https://doi.org/10.1016/S0031-9422(02)00137-1

Winter H, Huber SC (2000) Regulation of sucrose metabolism in higher plants: localization and regulation of activity of key enzymes. Crit Rev Biochem Mol Biol 35(4):253–289. https://doi.org/10.1080/10409230008984165

Yang L, Chen JH, Xu T, Zhou AS, Yang HK (2012) Rice protein improves oxidative stress by regulating glutathione metabolism and attenuating oxidative damage to lipids and proteins in rats. Life Sci 91(11–12):389–394. https://doi.org/10.1016/j.lfs.2012.08.003

Zaimoglu Z, Koksal N, Basci N, Kesici M, Gulen H, Budak F (2011) Antioxidative enzyme activities in Brassica juncea L. and Brassica oleracea L. plants under chromium stress. J Food Agric Environ 9(1):676–679

Zenk MH (1996) Heavy metal detoxification in higher plants—a review. Gene 179(1):21–30. https://doi.org/10.1016/S0378-1119(96)00422-2

Zhai R, Su S, Lu X et al (2005) Proteomic profiling in the sera of workers occupationally exposed to arsenic and lead: identification of potential biomarkers. Biometals 18(6):603–613. https://doi.org/10.1007/s10534-005-3001-x

Zhang M, Senoura T, Yang X, Nishizawa NK (2011) Functional analysis of metal tolerance proteins isolated from Zn/Cd hyperaccumulating ecotype and non-hyperaccumulating ecotype of *Sedum alfredii* Hance. FEBS Lett 585(16):2604–2609. https://doi.org/10.1016/j.febslet.2011.07.013

Zhang S, Cui Y, Li L et al (2015) Preparative HSCCC isolation of phloroglucinolysis products from grape seed polymeric proanthocyanidins as new powerful antioxidants. Food Chem 188:422–429. https://doi.org/10.1016/j.foodchem.2015.05.030

Environmental Lead and Wild Birds: A Review

Robert J. Williams, Steven D. Holladay, Susan M. Williams, and Robert M. Gogal, Jr.

Contents

R.J. Williams • S.D. Holladay • R.M. Gogal, Jr. (✉)
Department of Veterinary Biosciences and Diagnostic Imagining, College of Veterinary Medicine, University of Georgia, Athens, GA, USA
e-mail: rjowers@uga.edu; sdholl@uga.edu; rgogal@uga.edu

S.M. Williams
Poultry Diagnostic and Research Center, College of Veterinary Medicine, University of Georgia, Athens, GA, USA
e-mail: smwillia@uga.edu

P. de Voogt (ed.), *Reviews of Environmental Contamination and Toxicology Volume 245*, Reviews of Environmental Contamination and Toxicology 245, DOI 10.1007/398_2017_9

1 Introduction

1.1 Lead Composition, Classification, and Use

Elemental lead has an atomic mass of 207.2 g and density 11.34 g/cm^3. It is classified as a heavy metal by virtue of its high molecular weight and density $\geq$5.0 g/cm^3 (Tchounwou et al. 2012). As with other heavy metals, lead is considered to be a trace element and in the United States under natural conditions its mean environmental concentration in soil is 16 ppm (Shacklette and Boerngen 1984). Pure lead does not naturally exist in the environment, but rather it is found as a component of mineral ores. The most abundant of these ores is galena, which contain other elements that predominantly include silver (Dube 2006).

The first human documented use of lead dates back to the Roman Empire, after which lead use was minimal until societies became more industrialized (Milton 1988; Nriagu 1983). Lead use then dramatically increased such that, in 2015, more than 385,000 metric tons of lead were mined in the United States with an additional 1.12 million tons produced through recycling of postconsumer scrap (USGS 2016). These data verify that vast amounts of lead continue to be refined and used by consumers in the United States, and are then extended by even higher production in other countries such as China (USGS 2016). Based on the millions of tons of lead mined in the past and present years, it is not surprising that lead is viewed as a persistent global environmental threat. Further, while some environmental toxicants degrade over time, lead as a heavy metal element is relatively inert in its natural form.

1.2 Distribution and Biomarkers of Lead Toxicity

Systemic distribution of lead following exposure via inhalation or consumption occurs primarily through the circulatory system. Based largely on mammalian studies, it has been estimated that 30% of consumed lead and 50% of inhaled lead are absorbed into circulation (Agrawal 2012). Once in circulation, ~5% of total lead remains in blood while up to 95% distributes to other tissues, and is then slowly excreted from all these compartments through urine or feces (Agrawal 2012). Lead deposition in bone is highly persistent, occurs via the formation of a lead–phosphate complex, and can account for up to 95% of the total body burden from sub-chronic to chronic exposures (Agrawal 2012). The half-life of lead in the blood of mammals has been estimated as ~30 days while lead deposited in bone has a half-life of 20–30 years, with risk of lead reentering circulation and causing toxicities long after the initial exposure (Rabinowitz 1991). While limited research exists on the half-life of lead in the blood of avian species, Fry and Maurer determined that the half-life of lead in blood of California condors was approximately 13 days.

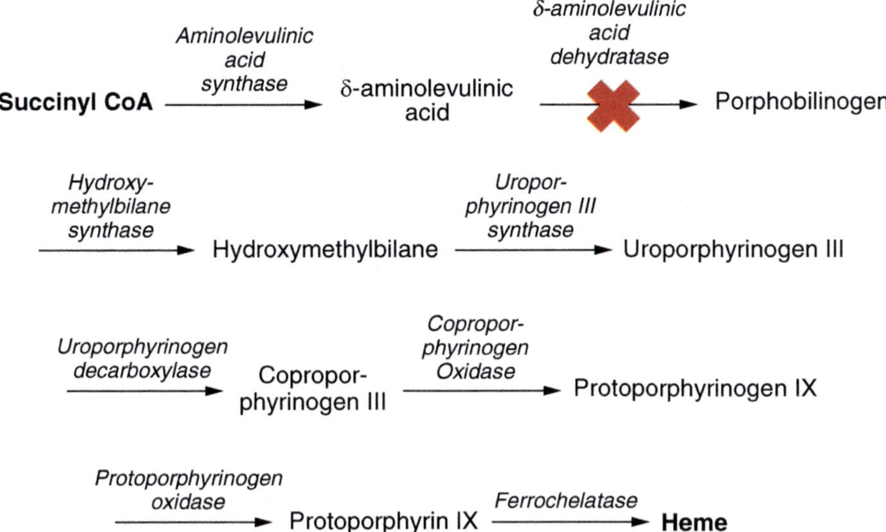

Fig. 1 Inhibition of heme synthesis by lead. The red X indicates the inhibition of heme synthesis through lead binding and suppressing δ-aminolevulinic acid dehydratase

Both blood and urine samples are commonly collected to determine the level of exposure to diverse toxicants as well as to help determine the severity of the systemic health effects to these toxic contaminants (Strimbu and Tavel 2010). The enzyme, delta aminolevulinic-acid dehydratase (δ-ALAD), is extremely sensitive to lead exposure, and decreased activity of this enzyme in blood cells is the classic biomarker of lead exposure (Berlin and Schaller 1974). δ-ALAD is the second of eight enzymes involved in heme synthesis; therefore, inhibition of this enzyme by lead significantly hinders an organism's ability to produce heme (Fig. 1). Another biomarker associated with lead toxicity is urine aminolevulinic acid (ALA). ALA is a substrate of δ-ALAD, so inhibition of the enzyme yields an increase in blood ALA concentrations. The increased ALA is filtered by the kidneys and is secreted into the urine. Both of these biomarkers are routinely used for assessing lead toxicity in human, domestic, and wildlife species as well as tracking responses to interventional treatment.

1.3 Cellular Mechanism of Lead Toxicity

All animal species depend on sodium, calcium, potassium, iron, and zinc for proper cellular function and survival. Lead, as a divalent cation, competes for ionic binding sites in proteins and thereby prevents essential divalent cations such as zinc and calcium from binding (Godwin 2001). This mechanism of competition and ion displacement by lead has been coined "ionic mimicry" (Kirberger and Yang 2008).

The biomarker protein for lead exposure, δ-ALAD, requires a single zinc ion for proper enzyme structure and function (Godwin 2001). The binding site for zinc in δ-ALAD contains three cysteine residues, to which lead binds with moderate affinity, slightly modifying the protein structure and inhibiting ALA from binding the active site (Godwin 2001).

Another class of high affinity receptors for lead is the calcium-dependent proteins such as calmodulin (Kirberger and Yang 2008). Calmodulin is a cytoplasmic protein expressed in nearly all cells of the body, which initiates a signal cascade for protein transcription following calcium binding at its four active sites (Wiemann et al. 1999; Wayman et al. 2011). Lead competes with calcium for calmodulin's four binding sites but also can bind at various points along the structure of the protein in an action known as "opportunistic binding" (Kirberger et al. 2013). The binding of lead to calmodulin initiates improper folding of the protein, resulting in inhibition of many cellular functions including prevention of signal transduction to the nucleus (Kirberger et al. 2013). Depending on a cell's level of dependence on calcium, the degree of lead's toxicity can thereby vary. This in part explains why some organ systems (i.e., nervous system) are more sensitive to lead exposure than other systems.

2 Lead Sources and Avian Species

2.1 Lead Ammunition and Avian Species

Avian wildlife encounter lead as a direct result of human sport shooting and hunting related activities. In particular, foraging avian species often consume particulate lead in the form of spent shotgun pellets or bullet fragments, confusing them for grit or food (Mateo et al. 2007; McConnell 1967; Schulz et al. 2002). It has been estimated during the twentieth century alone that approximately 3 million metric tons of lead were deposited in the form of spent ammunition in the United States (Craig et al. 1999). According to a 1999 estimate, approximately 60,000 additional tons of lead are deposited in high concentrations each year in the top few inches of soil at private and military shooting ranges as well as recreational hunting sites throughout the United States and other countries (Craig et al. 1999). According to Pain (1991) lead pellet distribution in Camargue, France (Rhone Delta) was as high as 2 million pellets/ha and over a 2-year period studying settlement of the pellets, concluding that the half-life of the pellets in the top 6 cm of soil was 46 years. Therefore, not only are certain avian species at risk of exposure to this unique form of lead, but they are also in a unique situation of elevated risk for multiple lead pellet or fragment exposures in these select sites.

2.2 *Lead Exposures from Sources Other than Lead Ammunition*

Although lead ammunition is a predominant source of lead pollution in the environment, wildlife species are also exposed to lead from other sources. Cai and Calisi (2016) examined 825 blood-lead level records from visibly ill feral pigeons in different neighborhoods of New York City collected during 2011–2015, a city where lead ammunition would likely be a minimal source of environmental lead. Their data showed that the area of the city where each pigeon was found did not influence the blood-lead levels; however, the blood-lead levels in the pigeons were higher in summer months compared to any other time of the year (Cai and Calisi 2016). The most interesting finding was that increased blood-lead levels in pigeons during summer months positively correlated with the blood-lead levels >10 μg/dL of children, suggesting that continued monitoring of feral pigeons in New York City for lead could help determine the risk for further human and wildlife exposure as well as guide efforts to decrease it (Cai and Calisi 2016).

Common sources of lead contamination for avian species are lead tackle and fishing weights. Sidor et al. (2003) analyzed 522 common loons (*Gavia immer*) found dead in New England from 1987 to 2000. The authors determined that 50% of the adult loons had died from lead toxicity induced by consumption of lead fishing tackle (Sidor et al. 2003). A similar case study examined three loons found dead in the Northeastern United States. Two of the three birds had fishing line sections with lead weights still attached in their ventriculus, while the third loon had a fishing line with no weights (Locke et al. 1982). All three loons had significantly increased lead concentrations in the liver and death was attributed to lead toxicosis (Locke et al. 1982). A similar study in Canada investigated the death of six loons found on the shore of a lake (Daoust et al. 1998). The authors reported lead fragments from sinkers in the ventriculus of 4 of the 6 loons and all six had significantly increased lead concentrations in the kidneys that were comparable to known lead poisoning concentrations (Daoust et al. 1998). Other wild aquatic avian species have been found with lead fishing tackle in their ventriculus including brown pelicans (*Pelecanus occidentalis*), Trumpeter swan (*Cygnus buccinator*), and a number of duck species (Locke et al. 1982; USEPA 1994; Franson et al. 2003).

Another source that contributes to heavy metal environmental pollution including lead is mining operations. In 2013 Beyer et al. examined the songbird populations in Missouri near a lead mining operation. It was already documented that the lead mining had yielded highly increased lead concentrations in the soil surrounding the facility. The rationale for the study was that bird populations consume earthworms living in the contaminated soil. The researchers showed that the songbirds in this region had lead levels increased by a factor of 8, 13, and 23 in the blood, liver, and kidney, respectively, as compared with songbirds in uncontaminated regions of Missouri (Beyer et al. 2013). Further, the δ-ALAD activity in these birds was decreased by approximately 50–80% (Beyer et al. 2013). Based on these results, the authors concluded that the health of bird populations living in areas near the mines was at high exposure risk to the lead and other heavy metals being extracted. As with mining operations, industrial smelting releases numerous metals into the environment through emissions. In Finland, liver accumulation of heavy metals including lead in passerine birds was reduced by 58–95% following reduction of atmospheric emissions from the country's largest nonferrous smelter (Berglund et al. 2012). Mallards given a commercial pellet diet containing 24% sediment from the contaminated Coeur d'Alene River Basin in Idaho (3,400 μg lead/g of sediment) had increased breast muscle atrophy, viscous bile, and increase in nuclear inclusion bodies in renal tubular cells (Heinz et al. 1999).

Another study attempted to link bird exposure to lead by way of natural lead concentrations in the environment rather than anthropogenic sources. The authors used the Alaskan tundra swan as the model for their experiment based on their migratory patterns in relation to lead contaminated sites (Ely and Franson 2014). Previously, it had been shown that, because of their winter migration patterns, Alaskan tundra swans are exposed to lead in the warmer climates where lead is prevalent due to human use (Ely and Franson 2014). However, during the summer months these swan populations migrate to northern areas of Alaska where there is very limited human influence. The researchers showed that during the summer months, the swan populations had on average <0.2 mg/mL blood-lead levels compared to winter months where birds had lead levels ranging from 0.2 to 0.5 mg/mL (Ely and Franson 2014). These results strongly suggest that the lead levels seen in avian tundra swan populations are less likely to be attributed to natural occurrences in their environment. Whooper swans wintering in Great Britain with blood-lead levels ≥4,400 mg/mL had significantly decreased winter body conditions, showing that sublethal levels of lead had a very detrimental impact on the health of the swans during winter months (Newth et al. 2016). These data coupled with the data from the swans in Alaska suggest that migratory birds exposed to anthropogenic lead during summer and winter months could experience adverse health consequences, even at sublethal doses of lead.

2.3 *Lead and Terrestrial Avian Species*

Much of the earlier studies addressing potential lead toxicity in birds were focused on aquatic species; however, terrestrial species are also at risk. Kerr et al. (2010) reported that northern bobwhite quail orally gavaged with five lead pellets (~45 mg/pellet) survived the acute exposure resulted in blood-lead levels that would be lethal to most mammalian species. Reasons for which avian species are more tolerant to higher lead are not currently known, however, may be linked to rapid ability to replenish their RBC and the fact that avian RBCs are nucleated. Kerr et al. (2010) suggested that, while these adult birds appear capable of surviving acute and possible chronic very high levels of lead, the long-term impact of such environmental lead exposure on avian reproduction and development remains unclear.

More recently, Holladay et al. (2012) orally gavaged Roller pigeons with 1, 2, or 3 lead pellets (45 mg/pellet). The Roller pigeons showed a higher tolerance to lead as compared to the Northern Bobwhite quail (Kerr et al. 2010), with no outward clinical signs at blood-lead levels up to 80 times higher than the control pigeons. The Roller pigeon has a higher tolerance to lead at these concentrations likely in part, due to its larger body mass compared to the bobwhite quail. Another noteworthy finding from this study was that lead pellet retention within the digestive tract varied between the two species. In the pigeons, mean pellet retention in the gut extended 4 weeks into the study whereas in the quail pellet mean retention time was less than 2 weeks (Holladay et al. 2012). A prolonged retention time of lead pellets by some species would presumably result in greater deposition of mobilized lead into the tissues of the birds and therefore pose a different risk of toxicity. Although not yet considered, prolonged retention time of lead pellets in some avian species might also result in increased risk of lead toxicity in predator or scavenger species of those birds. Another factor that could influence the variation in lead toxicity observed among avian species is the rate of uptake of lead through the ventriculous and distal digestive track. Kerr et al. (2010) working with quail and Holladay et al. (2012) with Roller pigeons recorded different retention times of the lead pellets between the two species, but did not assess whether the pellets were broken down, absorbed, or excreted. The observed increased mortality of the quail at the same dose of lead compared to the pigeon could be explained by the quail having a higher uptake of lead and therefore would have a higher acute toxicity.

A study by Mateo et al. (2001) in Spain found that approximately 11% (12/109 eagles) of the Spanish imperial eagle population, listed as an endangered species, had lead pellets in their digestive tract. The study also reported that the lead concentration in these eagles and the number of eagles containing lead increased during waterfowl hunting season (Mateo et al. 2001). These data suggest that waterfowl exposed to lead via oral consumption or nonlethal birdshot injury can be consumed by the eagles and thus act as a vector for lead exposure up the trophic levels. These data are not unique to the Spanish imperial eagle. A study by Cruz-Martinez et al. (2012) showed increased lead levels in the bald eagle in the United States from secondary exposure. These authors analyzed more than 1,200 eagles in Minnesota, and found that greater than 25% of the birds had elevated lead levels that were temporally associated with consuming of deer killed by hunters (Cruz-Martinez et al. 2012). Just as with the eagles of Spain, maximally increased lead levels were seen during hunting season and linked to lead ammunition. In addition, these researchers reported that the eagles had elevated copper concentrations (Cruz-Martinez et al. 2012). Lead ammunition is generally coated with a solid copper jacket; therefore the co-presence of lead and copper further supported indirect lead exposure through consumption of deer killed by hunters. Another case of lead exposure through scavenging was reported in the Iberian Peninsula where three lethargic Griffon vultures (*Gyps fulvus*) were found in the wild and died within 24 h of being transported to a clinic (Dvm et al. 2016). All three vultures were shown to have blood-lead levels ranging from 969 to 1,384 μg/dL, liver-lead levels ranging from 309 to 1,077 μg/dL dry weight, and one vulture was necropsied and had nine lead pellets in its stomach (Dvm et al. 2016).

The most publicized and likely most recognized case of secondary exposure of avian species to lead involves the California condor (*Gymnogyps californianus*). This high trophic predator has been on the endangered list since the 1980s when populations dwindled to 22 birds, but through intensive human management rebounded to approximately 400 birds by 2010 (Snyder and Snyder 2000; Walters et al. 2010). A recent survey by the United States Fish and Wildlife Service (USFWS) in 2015 reported the overall population to be at 435 (265 in the wild and 167 captive). A survey study of the California condors between 1997 and 2010 showed that approximately 50–88% of free flying condors had blood-lead levels exceeding the 100 ng/mL maximum safe level set in place by conservationists (Finkelstein et al. 2012). Approximately 20% of these same birds had blood-lead levels ≥450 ng/mL, which is the threshold set by the Center of Disease Control and Prevention for requiring immediate therapy in children (Finkelstein et al. 2012). In an attempt to alleviate the burden of lead exposure on the California condor and other wildlife, the California state government passed a law in 2013 banning the purchase and use of lead ammunition for hunting by 2019 (California Department of Fish and Wildlife 2017). A recent study examining the risk of lead exposure in condors reintroduced in California monitored blood-lead levels of these birds from 1997 to 2011 (Kelly et al. 2014). Even with the regulations on lead ammunition during waterfowl hunting season prior to the complete ban in 2013 the authors showed that the blood-lead levels of the reintroduced birds were significantly higher than those of birds still in captivity and the blood-lead levels increased with the age of the condor (Kelly et al. 2014). While this action prevents new lead from being deposited into soil in the form of spent ammunition, it does not impact previous deposited lead. The long-term outcome of this ban and the effect it will have on California condor won't be known for many years, again, due to the longevity of preexisting lead in the environment.

2.4 Lead and Aquatic Avian Species

Aquatic avian species such as waterfowl are susceptible to ingestion of spent lead ammunition much like other foraging avian species. One such case of lead exposure in aquatic birds occurred within a wildlife management area (WMA) in Kansas in 1979. Workers at the WMA found a total of 79 deceased Canadian geese plus 10 additional geese that were too lethargic to fly (Howard and Penumarthy 1979). Seventeen of the deceased geese were necropsied and were found to have an average of 13 lead pellets deposited in the crop and ventriculus, high lead concentrations in the kidney and liver, and other clinical signs of lead toxicity (Howard and Penumarthy 1979). In 2011–2012 in Argentina, 415 ducks (hunter killed and live captured) were analyzed for lead exposure (Ferrayra et al. 2014). While only 10% of the ducks had lead pellets in the digestive tracts, all 415 ducks had detectable bone-lead levels (Ferrayra et al. 2014). Although the bone-lead levels were not representative of lead toxicity, the presence of lead in the bone indicates that all birds were exposed to environmental lead.

Chronic lead exposure in waterfowl throughout much of the twentieth century coupled with increased lead deposition in bodies of water during waterfowl hunting season led to a nationwide ban on lead ammunition for waterfowl in 1991. This was done to help prevent further environmental contamination and loss of waterfowl populations from lead toxicity; however, exposure still occurs through ingestion of spent lead ammunition legally used in recreational shooting sites and for hunting of other wildlife. Romano et al. (2016) examined wetland areas and rice field for lead pellet contamination and found a varying degrees of pollution with areas having anywhere from 5.5 to 141 pellets/square meter. It was determined that in the areas with higher lead contamination, the water was slightly acidic indicating oxidation of lead and readily mobilization of lead into the waterways (Romano et al. 2016). California was one of the first states to implement replacing of lead pellets with steel pellets. Nonetheless, a study conducted 1 year after the removal of lead shot in California found that approximately 20% of the mallards captured at two separate wildlife refuges had elevated blood-lead levels (Mauser et al. 1990). Ten of the captured mallards died from what appeared to be lead toxicity and were found to have as few as one lead pellet in the ventriculus; while one bird had 52 lead pellets in its ventriculus at the time of death (Mauser et al. 1990). One study conducted in 2011–2012 in Spain on the Ebro Delta showed that even though lead shot had been banned since 2003 (thus 8–9 years earlier), approximately 16% of the mallard duck population had ingested lead in the form of spent shot (Vallverdu-Coll et al. 2015a). These findings are comparable to an Argentinian study where, of the 415 Mallard ducks sampled in 2012, 10.4% had ingested lead in the form of spent ammunition at the time of analysis (Ferrayra et al. 2014). Lead was detected in 100% of the bones analyzed from these same mallards, and in 60% of the livers (Ferrayra et al. 2014).

A study in Poland that evaluated both mallards and coots collected by hunters during 2006–2008 provided evidence of lead toxicity in both species (Binkowski et al. 2013). Histological analysis showed inflammatory and leukocytic infiltrative lesions consistent with lead toxicity in the birds, with the liver and kidney most affected (Binkowski et al. 2013). These data again show that lead exposure in aquatic birds across diverse locations remains a critical problem. As might be expected, additional studies are showing that lead toxicity is not a problem limited to waterfowl, but also can affect other aquatic avian species, such as marsh and coastal birds. For example, the black-necked stilt is a bird primarily found in the marshes of Texas. A recent study conducted in the Upper Texas Coast found that almost 80% of the test population had blood-lead levels exceeding 20 μg/dL (Riecke et al. 2015). Although these blood-lead levels were low compared to other studies, they could still potentially have long-term adverse outcomes on bird population health (Riecke et al. 2015).

Although growing focus is being given to health effects of lead in avian wildlife species, the co-presence of other environmental contaminants must be considered to truly estimate risk to different bird population health. Two recent studies reported increased levels of mercury, lead, arsenic, cadmium, chromium, and other metals in two geographically different locations in the United States (Burger et al. 2014; St Clair et al. 2015). The avian species analyzed in these experiments were the Semipalmated sandpipers in Delaware Bay and the Pacific Dunlin on the west coast (Burger et al. 2014; St Clair et al. 2015). These studies show that not just environmental lead contamination, but heavy metal contamination in general remains a nationwide issue.

3 Avian Lead Toxicities

3.1 Organ System Toxicity

Lead is a toxicant that has the ability to affect multiple organs and organ systems in numerous species, including avian species. The classic toxicity seen in all animal models is the depression of delta-aminolevulinic acid dehydratase (δ-ALAD). This hematologic toxicity has been documented for decades and is so specific to lead that it is routinely used as a biomarker to verify lead exposure (Pain 1989; Binkowski and Sawick-Kapusta 2015). Martinez-Haro et al. (2011) showed that mallards and coots with blood-lead levels as little as 6 μg/dL had decreased δ-ALAD activity. Lead in the bloodstream has two fates, excretion or deposition into tissue.

The liver is the largest soft tissue store of lead in the body and is therefore commonly analyzed in avian lead toxicity cases and studies (Aloupi et al. 2017; Behmke et al. 2017). Mateo et al. (2003) reported that mallards exposed to dietary lead, 2 g/kg, for a period of 3 weeks had increased lipid peroxidation in the blood, liver, and bile. There was also a negative correlation with lipid peroxidation and glutathione peroxidase activity in the liver (Mateo et al. 2003). A similar study using Japanese quail showed that a single exposure to lead shot via oral gavage resulted in significantly reduced glutathione activity coupled with increased lipid peroxidation (Osickova et al. 2014). These results have been shown in diverse avian species studied including vultures, pied flycatcher (*Ficedula hypoleuca*), Canada geese (*Branta canadenisis*), and many more (Mateo and Hoffman 2001; Berglund et al. 2007; Behmke et al. 2017).

The renal system is another target of lead toxicity. Numerous laboratory studies have used acid-fast staining to show inclusion bodies in the nuclei of renal cells after relatively high levels of lead exposure (Locke et al. 1966; Kelly et al. 1998). It has also been shown that subchronic to chronic exposures to lead at lower doses can also result in these inclusion bodies. For instance, a team of researchers in the United Kingdom investigated an urban population of pigeons (*Columba livia*) and found that the majority of birds necropsied had significantly increased nuclear inclusion bodies, all of which contained lead (Johnson et al. 1982). Further, it was determined that the lead within the kidneys was trialkyl lead, most commonly formed from the exhaust of vehicles using fuel with lead additives (Johnson et al. 1982). Another group of researchers verified that kidneys from mallards exposed to lead via contaminated sediment not only contained these classic inclusion bodies, but also had decreased glutathione activity and increased oxidative stress in renal cells (Hoffman et al. 2006). Kidney toxicosis from lead was seen in red-crowned cranes (*Grus japonensis*) in northern China indicating lead exposure is a serious threat to the endangered species (Luo et al. 2016).

Lead exposure, under select conditions, has also been found to alter immune function of avian species and can increase susceptibility to infection. Birds do not have lymph nodes like mammals and therefore the spleen is more critical to the integrity of the immune system. Ferreyra et al. (2015) reported that wild mallards from wetlands in Argentina with known high lead contamination had significantly lower spleen/bodyweight ratios compared to mallards from other areas. These same birds showed a negative correlation between bone-lead concentrations and spleen size. Another study monitored a wild population of great tit (*Parus major*) nestlings in a known metal contaminated site, for four innate immune markers: agglutination, lysis, haptoglobin concentrations, and nitric oxide concentrations (Vermeulen et al. 2015). These researchers showed that among birds displaying the greatest signs of lead toxicity, the innate immune functional marker most altered was the capacity to induce cell lysis. A similar study evaluated the transfer of maternal lead to eggs and offspring, including characterizing lead-induced alterations to the developing immune system (Vallverdu-Coll et al. 2015b). These researchers collected wild mallard eggs from known lead contaminated areas and measured eggshell lead concentrations and duckling blood-lead levels to determine exposure. They then showed a negative correlation with blood-lead levels and the T cell mediated immune response measured with the PHA skin test. After hatching, the F_1 ducklings also had decreased antioxidant activity and increased reactive oxygen species with increased blood-lead levels (Vallverdu-Coll et al. 2015b). Dietary lead exposure in adult chickens revealed that after 30- and 60-day exposure, peripheral blood lymphocytes had an increase in mRNA expression of the pro-inflammatory cytokines tumor necrosis factor-α (TNF-a), cyclooxygenase-2 (COX-2), and inducible nitric oxide synthase (iNOS) (Sun et al. 2016). This would suggest that the circulating lymphocytes were being activated by lead in the blood. The implications from this research may be particularly critical for future lead research in avian species, because it would imply that maternal transfer of lead to offspring in the wild may adversely impact the developing immune system of the avian offspring. This in turn may suggest, in areas where lead contamination is prevalent, that entire populations could be at elevated risk for disease susceptibility. Another team of researchers examined the role of dietary lead exposure and gene expression of HSP (27, 40, 60, 70, and 90) and inflammatory factors iNOS, tumor necrosis factor-α TNF-α, and COX-2 in the cartilage of chickens (Zheng et al. 2016). The gene expression of all the HSP and inflammatory factors were increased through lead exposure, suggesting lead was causing an inflammatory response in the cartilage and providing evidence that HSP could be used as a biomarker of lead-induced damage (Zheng et al. 2016).

As with mammalian species, avian species are susceptible to neurotoxic effects induced by lead. Burger and Gochfeld (1994) injected 1–2-day-old herring gulls intraperitoneally with 100 mg/kg lead acetate and observed behavior, activity, feeding habits, and general health. Approximately 65% of the lead injected birds survived to day 14 compared to 96% of the control birds; of the surviving lead injected birds, significant neurologic deficits were observed through increased stumbling when walking, decreased begging for food, and decreased accuracy when pecking the parent bird's bills for feeding (Burger and Gochfeld 1994). Douglas-Stroebel et al. (2004) reported that mallard ducklings given feed containing 24% sediment from the Coeur d'Alene River Basin, a known contaminated river basin with lead levels reaching 3,449 μg/g, had brain-lead levels 485 ppb. These ducklings were also reported to have lower brain weights and increased oxidized glutathione as compared to the control ducklings (Douglas-Stroebel et al. 2004). The ducklings in the 24% river basin sediment treatment group also had behavioral changes including decreased time swimming and issues with balance and mobility (Douglas-Stroebel et al. 2005). American kestrel (*Falco sparverius*) nestlings exposed to 25, 125, and 625 mg/kg metallic lead in corn oil had decreased brain RNA to protein ratios and brain δ-ALAD, while brain monoamine oxidase and ATPase were not significantly different (Hoffman et al. 1985). These biochemical changes were more severe in the nestlings compared to adults or young precocial birds exposed to lead (Hoffman et al. 1985).

3.2 Reproductive Toxicity

Another area of research that has gained increased attention in the past few decades is lead-induced reproductive toxicity in avian species. It has previously been shown that hens from multiple species, exposed to lead in the environment or laboratory settings, deposit lead into the eggshells and egg contents during lay periods (Finley et al. 1976; Burger 1994; Trampell et al. 2003; Burger and Gochfeld 2004; Tsipoura et al. 2011). Vallverdu-Coll et al. (2015b) showed that developmental exposure to lead in red-legged partridges caused sex-specific and seasonal changes. In particular, the male birds showed increased coloration during spring months, while females showed increased humoral antioxidant activity. The researchers concluded that the males most likely increased coloration to increase chances of breeding while the females increased antioxidant levels to defend against reactive oxygen species commonly produced by lead.

Veit et al. (1983) orally gavaged ringed turtledoves with four lead pellets (110 mg/pellet) and examined testicular changes. The birds treated with lead had significantly lower testes weight and, histologically, displayed significant degeneration suggesting suppressed reproductive capabilities (Veit et al. 1983). Another study examining lead toxicity showed that male red-legged partridges exposed to lead shot had altered sperm quality and decreased motility (Vallverdú-Coll et al. 2016). An increase in heat shock proteins (HSP) has been correlated with lead-induced cell damage in avian testis. Huang et al. (2017) showed that selenium administered with lead in the water of Hyline chickens prevented testicular damage and decreased mRNA expression of HSP70. The researchers concluded that HSP70 could be a biomarker of testicular damage for lead in the avian model (Huang et al. 2017).

Vallverdu-Coll et al. (2015a) reported that mallard hatchlings from the Ebro Delta of Spain with blood-lead concentrations greater than 180 ng/mL died within 1 week post hatch. These hatchlings were collected as eggs and were not exposed to lead in the laboratory setting; therefore these blood-lead levels were a result of maternal deposition. Another study conducted by the same group evaluated red-legged partridge (*Alectoris* rufa) hens gavaged with three #6 Pb pellets (~109 mg/pellet), and showed a decrease in hatchability of eggs laid (Vallverdú-Coll et al. 2016).

Ruiz et al. (2016) examined dietary lead exposure on great tit nestlings and the role of vitamins A, D3, K, and E, which are essential for growth and development. The researchers concluded that the nestlings retained higher concentrations of plasma vitamin A in an attempt to counter the stress induced by lead during growth and development (Ruiz et al. 2016). Sediment contaminated with lead is another dietary exposure in avian species (Beyer et al. 2000; Hoffman et al. 2000a). Hoffman et al. (2000a) fed day old mallard ducklings feed mixed with 0, 12, and 24% sediment diet (3,449 μg lead/g sediment) from the Coeur d'Alene River basin. At 6 weeks of age, the ducklings in the 24% diet had liver and kidney lead concentrations of 7.92 and 7.97 ppm, respectively, and a 40% reduction of hepatic glutathione and nuclear inclusion bodies in renal cells (Hoffman et al. 2000a). These results were duplicated in Canada geese receiving an identical diet with an additional group receiving a 48% diet (Hoffman et al. 2000b). The geese receiving the 48% diet had liver and kidney lead concentrations of 6.57 and 14.93 ppm, respectively, with one gosling suffering from subacute renal tubular nephrosis (Hoffman et al. 2000b). The collective available results therefore suggest that both hens and cocks of avian species are susceptible to gender-specific reproductive toxicities after exposure to lead. Further, maternal lead can be deposited in the eggs, with resulting effects on the development and survival of the hatchlings.

One thing to note is that hatchlings of altricial species such as passerines and raptors are more sensitive to lead exposure than hatchlings of precocial species since they are less physiologically advanced at hatching and thus totally dependent on parents for feeding similar to mammalian neonates. Thus, this may constitute one of the most sensitive periods of the avian life cycle with respect to lead and other environmental contaminants (Hoffman et al. 2003).

4 Diagnostic and Therapeutic Protocols for Birds with Elevated Blood-Lead Levels

The least invasive technique for diagnosing lead exposure in avian species is radiographic imaging. Radiographs can be performed in at risk avian species such as the condor to determine if radio-dense lead is present in the crop or stomach

(Redig and Arent 2008). In clinical avian cases where orally consumed particulate lead is detected, it can be physically removed from the crop and/or stomach, and if lead is located more distally in the digestive tract then treatment to induce excretion is employed (Redig and Arent 2008). The caveat with radiographs is that if the pellet has already been metabolized or excreted, then it will not be visible and diagnosis of lead toxicity can only be determined through blood analysis.

The current and most effective treatment for elevated blood-lead levels is chelation therapy. Chelation therapy is used in human medicine to treat a number of issues including metal toxicities such as lead, arsenic, and mercury. Selection of an effective chelator requires that the chemical meet certain criteria. These include the chemical's ability to cross biologic barriers to reach metals, effective binding to metals and formation of stable compounds, and upon binding, rendering the metal nontoxic and increasing the excretion from the body (Jones 1994). The common chemicals used for chelation therapy include calcium disodium ethylenediaminetetraacetic acid (calcium EDTA), dimercaptosuccinic acid (DMSA), and dimercaprol (Garcia and Snodgrass 2012). These chemicals are injected directly into the blood stream of exposed individuals and bind lead to prevent further deposition into tissues and bone, and facilitate metal excretion (Garcia and Snodgrass 2012). Treatment generally involves a combination of the chemicals with dimercaprol being administered first and calcium EDTA following (Garcia and Snodgrass 2012). Chelation therapy is a multiple step process given over a number of days to ensure that blood-lead levels decrease to normal; follow up is also required because chelation therapy does not remove deposited lead from the body. Thus, deposited lead from tissue or bone can reenter the blood stream and cause further toxicities necessitating additional treatment (Garcia and Snodgrass 2012).

Chelation therapy has been used to treat birds with high blood-lead levels. Monitored free flying California condors and other endangered avian species with blood-lead levels of 450 ng/mL have been treated using chelation therapy (Finkelstein et al. 2012). The blood-lead level threshold of 450 ng/mL in birds is equivalent to blood-lead level of 45 μg/dL in humans, the value set as an action level for treatment in children (CDC 2012). The fact that condors have reached blood-lead levels that require chelation therapy shows how dire the situation is for these highly endangered species. The most common chelator used in avian species has been EDTA. Use of EDTA in humans and other mammals has been shown to be toxic in some cases where increased concentrations of EDTA caused nephrosis and central nervous system toxicity; however, none of these toxicities have been recorded in avian species, including when using higher concentrations of EDTA over prolonged periods of time (Redig and Arent 2008; Samour and Naldo 2002).

Diagnosing lead poisoning in wild birds can be challenging, with delivery of test results for blood-lead levels usually requiring a number of days to be returned. One case in Minnesota involved a wild Golden eagle, which lacked the ability to fly or properly perch (Shimmel and Snell 1999). Initially the clinicians diagnosed the eagle as having been exposed to a cholinesterase inhibitor insecticide due to its behavior and leg paralysis and immediately initiated treatment; it wasn't until after 4 days posttreatment during which the bird showed no clinical improvement, that the clinicians began intramuscular EDTA treatment for lead poisoning (Shimmel and Snell 1999). Within 24 h post-EDTA treatment, the eagle started to regain toe movement; however, it wasn't until after 6 weeks of intramuscular and oral EDTA treatment that blood-lead levels returned to reference intervals (Shimmel and Snell 1999). The original misdiagnosis was due to lead not being visible in radiographs, which is a common hurdle clinicians face.

Another case in New Zealand involved an Australian harrier found on the side of the road and delivered to an avian rescue center, with the belief that the bird had been hit by a car (Nijman 2016). Just as with the case of the above-described Golden eagle, no lead was apparent in the radiographs; however, the harrier was lethargic and blood-lead levels were 0.65 mg/L, representing the upper limit of the Leadcare Analyzer used by the attending veterinarian (Nijman 2016). Following immediate intramuscular administration of CaEDTA and days of oral chelation therapy, blood-lead levels returned to reference levels and the harrier regained energy and mobility (Nijman 2016). The challenge in both cases was that the lead was systemic and therefore not visualized as particles in the radiographs. The extended treatment using EDTA demonstrates the difficulty in removing sequestered lead from birds as well as the amount of human effort and time required to treat a bird suffering from lead toxicity. Although the cost for treating these birds was not provided, the treatment duration would suggest it to be considerable. Further, these cases show that even with a thorough medical workup, it is difficult to determine when and how often an individual bird may have been exposed to lead.

A bird's nutritional status is also an important factor that can contribute to lead reentering the blood via the tissues. Mobilization of lead and other elements from bone can occur when nutrient demand is high and dietary intake is inadequate. This may have been a contributing factor in the above case involving the clinically ill Golden eagle whose blood-lead levels never dropped below 0.08 ppm following 4 weeks of EDTA treatment. The eagle had a recorded dietary intake deficiency accompanying the infection (Shimmel and Snell 1999). Thus, this form of semi-chronic or chronic lead exposure, as a result of lead mobilization from the tissues, could be an important additional health hazard to avian species such as the endangered condors. Heinz et al. (1999) demonstrated in adult mallards that a total corn diet, which is a nutritionally inferior diet (Jordan 1968), increased the uptake or storage of lead from contaminated mining sediment in the liver. Lead contaminated sediment ingestion in mallard ducklings generally affected more variables in combination with a less optimal diet with respect to hematological, hepatic, and renal effects (Hoffman et al. 2000a). Also, Schueuhammer (1997) reported that the level of calcium (0.3% vs. 3.0%) in the diet had a significant effect on lead accumulation in liver and kidney of birds, with greater lead accumulation in the tissues of birds with low dietary calcium. Lead toxicity associated with ingested lead shot has also been shown to be greater in birds on a diet consisting primarily of corn, possibly due to corn's low total protein content and deficiencies in certain amino acids, calcium, phosphorus, and zinc, as well as an increased abrasion of lead pellets in the gizzard when corn is present (Jordan 1968; Finley et al. 1976; Carlson and Nielsen 1985).

5 Environmental Removal and Remediation

As previously stated, a major source of environmental lead is particulate from spent ammunition concentrated at outdoor shooting ranges, managed hunting areas, and military bases. This particulate lead is found in the top 1–3 in. of soil leaving foraging avian species and other animals at extremely high potential risk for consumption (Pain 1991). In an effort to prevent spent ammunition from contaminating the environment and poisoning local wildlife, the Environmental Protection Agency (EPA) published a guideline in 2005 titled *Best Management Practices for Lead at Outdoor Shooting Ranges*, which explains a series of management protocols for more sustainable and safe management of spent lead.

The first management technique recommended by the EPA was to create a containment area for the spent ammunition using "earthen backstops" or other traps to isolate the lead in a manageable area (EPA 2005). This approach was effective for target shooting; however, ranges that offer sport clay shooting do not utilize earthen traps, and thus the pellet dispersal (i.e., hundreds of lead pellets) is over an open field. To prevent the pellets from breaking down and leaching lead into the environment, the EPA has recommended monitoring the pH of the soil and keeping it at a more neutral pH (EPA 2005). Acidic environments break down the lead pellets and mobilize them into the soil or water nearby, thus increasing the potential exposure of more plant and animal species (EPA 2005). Adding phosphate to the soil prevents additional leaching of lead due to complexing of phosphate and lead to form a stable compound, decreasing the mobilization of lead (EPA 2005). At target and sport clay shooting ranges the most efficient technique for lead management is removal and recycling. Range owners can remove spent lead ammunition through raking and sifting as well as other manual techniques (EPA 2005). This is beneficial not only because it removes lead from the environment, but also because the spent ammunition can be washed and reused, essentially lowering the amount of lead needed for future ammunition production. At larger target ranges and sport clay ranges, outside third-party companies that specialize in lead ammunition removal can be hired (EPA 2005). Companies such as these remove the top several inches of soil and remove the lead using machinery prior to spreading the soil back on the range (EPA 2005). Again, in cases such as this, the lead can be recycled for the range manager or taken elsewhere by the company to be used. The final management technique issued by the EPA is keeping detailed logs of all shooting activity on the range (EPA 2005). Records of shooting activity allow for a more accurate estimate of how much lead has been deposited on the range compared to how much is reclaimed through recycling techniques.

In the case of California condor lead poisoning, outdoor shooting ranges were not the source of lead. Instead the California condors most likely consumed lead ammunition deposited in muscle tissue of game birds and mammals (Finkelstein et al. 2012). Hunting is a nationwide activity in the United States and the lands where hunting occurs are generally not managed like outdoor shooting ranges. Instead, deposited ammunition is predominantly left in the environment causing local contamination, and animals wounded or killed and not found leave carnivores and scavengers such as the California condor at risk of exposure. The most effective way to prevent lead from entering the environment through recreational hunting is to commercially remove lead as a component of ammunition. California has already begun the process of removing lead ammunition from use in all hunting, which will be completed in 2019; however, the remaining 49 states only require non-lead ammunition for waterfowl hunting. Until all 50 states require non-lead ammunition for hunting, wildlife will be at ongoing, likely increasing, risk of exposure to lead in the form of spent ammunition.

6 Conclusion

The physical characteristics and abundance of lead have made this heavy metal a useful component of building materials, cooking utensils, weaponry, plumbing, fuel additives, and many more products throughout Roman Empire and post-Roman Empire human history. The same characteristics that have attributed to the anthropologic use of lead contribute to lead being a persistent environmental toxicant (Battelle and EPA 1998; Dube 2006). As technology has advanced, the uses of lead have only grown, further compounding the potential of environmental pollution (USGS 2016). Over the last century, the multiple organ system toxicities induced by lead have been recognized and studied to quantify the level of exposure that will cause no detrimental effect; however, the overall outcome of this research has shown that there is no safe level of lead exposure.

The major effort in lead research has focused on human exposure and toxicity. This effort was driven by studies during the latter half of the twentieth century, which showed that lead leached into the environment from automobile emissions and that children could be exposed to lead via paint chips (McElvaine et al. 1992; Needleman 2000; Bridbord and Hanson 2009). Based on these past studies and the strict laws enacted, human lead exposure has been greatly reduced and reports of recent human lead exposure incidents such as the Flint Michigan incident (Craft-Blacksheare 2017; Sadler et al. 2017) are increasingly rare. Currently, the greatest threat of lead exposure is from the environment and the animals and plants that inhabit it (Table 1).

Although California has taken noteworthy steps towards protecting wildlife species, in particular the California condor, additional efforts are needed at the national level. It is currently unknown how many populations of different avian species across the United States are at risk of harm from lead exposure, or how diverse additional environmental factors may influence that risk. In addition, there are very little data on the effects of subclinical lead exposure on the long-term, trans-generational population health of avian species. This is an area that warrants further investigation. If the protection of all species is a goal at the state and federal level, then the removal of lead from ammunition is necessary. Further, more aggressive environmental remediation at sites where lead ammunition has been used in

Table 1 Avian species and lead exposure model

Species	Latin name	Exposure model	Reference
Northern pintail	*Anas acuta*	Environmental	Mateo et al. (2007)
Greylag goose	*Anser anser*	Environmental	Mateo et al. (2007)
Greater flamingo	*Phoenicopterus ruber*	Environmental	Mateo et al. (2007)
Glossy ibis	*Plegadis falcinellus*	Environmental	Mateo et al. (2007)
Mourning doves	*Zenaida macroura*	Environmental	Schulz et al. (2002)
Northern bobwhite	*Colinus virginianus*	Oral gavage 1, 5, or 10.45 mg pellets	Kerr et al. (2010)
Roller pigeon	*Columba livia*	Oral gavage 1, 2, or 3.45 mg pellets	Holladay et al. (2012)
Spanish emperial eagle	*Aquila adalberti*	Environmental	Mateo et al. (2007)
Red kite	*Milvus milvus*	Environmental	Mateo et al. (2007)
Bald eagle	*Haliaeetus leucocephalus*	Environmental	Cruz-Martinez et al. (2012)
Griffon vultures	*Gyps fulvus*	Environmental	Dvm et al. (2016)
California condor	*Gymnogyps californianus*	Environmental	Walters et al. (2010)
Canada geese	*Branta canadensis*	Environmental	Howard and Penumarthy (1979)
Whistling duck	*Dendrocygna bicolor*	Environmental	Ferrayra et al. (2014)
White-faced tree duck	*Dendrocygna viduata*	Environmental	Ferrayra et al. (2014)
Black-bellied whis-tling-duck	*Dendrocygna autumnalis*	Environmental	Ferrayra et al. (2014)
Rosy-billed pochard	*Netta peposaca*	Environmental	Ferrayra et al. (2014)
Brazilian duck	*Amazonetta brasiliensis*	Environmental	Ferrayra et al. (2014)
Mallard duck	*Anas platyrhynchos*	Environmental	Vallverdu-Coll et al. (2015a, b)
Coot	*Fulic ralidae*	Environmental	Binkowski et al. (2013)
Black-necked stilt	*Himantopus mexicanus*	Environmental	Riecke et al. (2015)
Semipalmated sandpiper	*Calidris pusilla*	Environmental	Burger et al. (2014)
Pacific dunlin	*Calidris alpina pacifica*	Environmental	St Clair et al. (2015)
Black duck	*Anas rubripes*	Environmental	Pain (1989)
Common pochard	*Aythya ferina*	Environmental	Martinez-Haro et al. (2011)
Black vultures	*Coragyps atratus*	Environmental	Behmke et al. (2017)
Turkey vultures	*Cathartes aura*	Environmental	Behmke et al. (2017)
Pied flycatcher	*Ficedula hypoleuca*	Environmental	Berglund et al. (2007)
Red-crowned cranes	*Grus japonensis*	Environmental	Luo et al. (2016)
Great tit	*Parus major*	Environmental	Vermeulen et al. (2015)
Red-legged partridge	*Alectoris rufa*	1 or 3,109 mg pellet oral gavage	Vallverdu-Coll et al. (2015a, b)
Broiler chicken	*Gallus gallus domesticus*	0.5 or 350 mg/kg dietary lead acetate	Sun et al. (2016)

(continued)

Table 1 (continued)

Species	Latin name	Exposure model	Reference
Ringed turtle dove	*Streptopelia risoria*	4,100 mg lead pellet oral gavage	Veit et al. (1983)
Common loon	*Gavia immer*	Environmental	Sidor et al. (2003)
Brown pelican	*Pelecanus occidentalis*	Environmental	Franson et al. (2003)
Double-crested cormorants	*Phalacrocorax auritus*	Environmental	Franson et al. (2003)
Black-crowned night herons	*Nycticorax nycticorax*	Environmental	Franson et al. (2003)
Northern cardinal	*Cardinalis cardinalis*	Environmental	Beyer et al. (2013)
American robin	*Turdus migratorius*	Environmental	Beyer et al. (2013)
Tundra swan	*Cygnus columbianus*	Environmental	Berglund et al. (2012)
Whooper swan	*Cygnus cygnus*	Environmental	Newth et al. (2016)
Golden eagle	*Aquila chrysaetos*	Environmental	Redig and Arent (2008)
Saker falcon	*Falco cherrug*	Environmental	Samour and Naldo (2002)
Peregrine falcon	*Falco peregrinus*	Environmental	Samour and Naldo (2002)
Lanner falcon	*Falco biarmicus*	Environmental	Samour and Naldo (2002)
Swamp harrier	*Circus approximans*	Environmental	Nijman (2016)

high quantities is crucial to prevent the leaching and spread of lead into surrounding areas. The breakdown of these bullets into smaller fragments increases the potential for consumption by animal species, especially avian species but also increases the potential for soil, water, and plant contamination, compounding the ability for lead to enter higher trophic levels.

7 Summary

This review summarizes historic and recent reports of lead toxicity in avian wildlife species. These reports show that both aquatic and terrestrial birds are at continued risk of lead exposure from sources that include spent lead ammunition, lead fishing weights, and industry-related contaminated lead sediments. Avian exposure to such lead at sublethal doses can cause detectable toxicities in multiple organ systems, with harmful reproductive and developmental effects being of increased recent concern. Diagnosis and treatment of lead toxicity are costly and generally reserved for only the most endangered of species, particularly California condors and several eagle species. Minimizing further exposures to these and other birds will most effectively be achieved by blocking entry of new lead into the environment, through measures such as California's ban on lead ammunition by 2019. Remediation techniques to remove existing lead from the environment will also help prevent exposure of wildlife; however, in most cases will be prohibitively expensive.

References

Agrawal A (2012) Toxicity and fate of heavy metals with particular reference to developing foetus. Adv Life Sci 2(2):29–38

Aloupi M, Karagianni A, Kazantzidis S, Akriotis T (2017) Heavy metals in liver and brain of waterfowl from the Evros delta, Greece. Arch Environ Contam Toxicol 72(2):215–234

Battelle and EPA (1998) Sources of lead in soil: a literature review. Battelle Memorial Institute and U.S. Environmental Protection Agency, Washington

Behmke S, Mazik P, Katzner T (2017) Assessing multi-tissue lead burdens in free-flying obligate scavengers in eastern North America. Environ Monit Assess 189(4):139

Berglund AM, Sturve J, Forlin L, Nyholm NE (2007) Oxidative stress in pied flycatcher (*Ficedula hypoleuca*) nestlings from metal contaminated environments in northern Sweden. Enrivon Res 105(3):330–339

Berglund AM, Rainio MJ, Eeva T (2012) Decreased metal accumulation in passerines as a result of reduced emissions. Environ Toxicol Chem 31(6):1317–1323

Berlin A, Schaller KH (1974) European standardized method for the determination of delta-aminolevulinic acid dehydratase activity in blood. Z Klin Chem Klin Biochem 12:389–390

Beyer WN, Audet DJ, Heinz GH, Hoffman DJ, Day D (2000) Relation of waterfowl poisoning to sediment lead concentrations in the Coeur d'Alene River Basin. Ecotoxicology 9(3):207–218

Beyer WN, Franson JC, French JB, May T, Rattner BA, Sheam-Bochsler VI, Warner SE, Weber J, Mosby D (2013) Toxic exposure of songbirds to lead in the southeast Missouri lead mining district. Arch Environ Contam Toxicol 65(3):598–610

Binkowski LJ, Sawick-Kapusta K (2015) Lead poisoning and its in vivo biomarkers in mallard and coot from two hunting activity areas in Poland. Chemosphere 127:101–108

Binkowski LJ, Starwarz RM, Zakrzewski M (2013) Concentrations of cadmium, copper, and zinc in tissues of mallard and coot from southern Poland. J Environ Sci Health B 48(5):410–415

Bridbord K, Hanson D (2009) A personal perspective on the initial federal health-based regulation to remove lead from gasoline. Environ Health Perspect 117(8):1195–1201

Burger J (1994) Heavy metals in avian eggshells: another excretion method. J Toxicol Environ Health 41:207–220

Burger J, Gochfeld M (1994) Behavioral impairments of lead-injected young herring gulls in nature. Fundam Appl Toxicol 23:553–561

Burger J, Gochfeld M (2004) Metal levels in eggs of common tern (*Sterna hirundol*) in New Jersey: temporal trends from 1971–2002. Environ Res 94:336–343

Burger J, Gochfield M, Niles L, Dey A, Jeitner C, Pittfield T, Tsipoura N (2014) Metals in tissues of migrant semipalmated sandpipers (*Calidris pusilla*) from Delaware Bay, New Jersey. Environ Res 133:362–370

Cai F, Calisi RM (2016) Seasons and neighborhoods of high lead toxicity in New York City: the feral pigeon as a bioindicator. Chemosphere 161:274–279

California Department of Fish and Wildlife (2017) Nonlead ammunition in California. https://www.wildlife.ca.gov/hunting/nonlead-ammunition. Accessed 25 May 2017

Carlson BL, Nielsen SW (1985) Influence of dietary calcium on lead poisoning in mallard ducks (Anas platyrhynchos). Am J Vet Res 46:276–282

Center for Disease Control and Prevention (CDC) (2012) Blood Lead Levels in Children. https://www.cdc.gov/nceh/lead/ACCLPP/blood_lead_levels.htm. Accessed 6 Oct 2017

Craft-Blacksheare MG (2017) Lessons learned from the crisis in Flint, Michigan regarding effects of contaminated water on maternal and child health. J Obstet Gynecol Neonatal Nurs 46 (2):258–266

Craig JR, Rimstidt JD, Bonnaffon CA, Collins TK, Scanlon PF (1999) Surface water transport of lead at shooting range. Bull Environ Toxicol 63:312–319

Cruz-Martinez L, Redig PT, Deen J (2012) Lead from spent ammunition: a source of exposure and poisoning in bald eagles. Human-Wildl Interact 6(1):94–104

Daoust PY, Conboy G, McBurney S, Burgess N (1998) Interactive mortality factors in common loons from maritime Canada. J Wildl Dis 34(3):524–531

Douglas-Stroebel E, Hoffman DJ, Brewer GL, Sileo L (2004) Effects of lead-contaminated sediment and nutrition on mallard duckling brain growth and biochemistry. Environ Pollut 131(2):215–222

Douglas-Stroebel E, Brewer GL, Hoffman DJ (2005) Effects of lead-contaminated sediment and nutrition on mallard duckling behavior and growth. J Toxic Environ Health A 68(2):113–128

Dube RK (2006) The extraction of lead from its ores by the iron-reduction process: a historical perspective. Archaotechnology feature. JOM 58:18–23

Dvm MA, Oliveria PA, Brandao R, Francisco ON, Velarde R, Lavin S, Colaco B (2016) Lead poisoning due to lead-pellet ingestion in Griffon vultures (*Gyps fulvus*) from the Iberian Peninsula. J Avian Surg 30(3):274–279

Ely CR, Franson JC (2014) Blood lead concentrations in Alaskan tundra swans: linking breeding and wintering areas with satellite telemetry. Ecotoxicology 23(3):349–356

Ferrayra H, Romano M, Beldomenico P, Caselli A, Correa A, Uhart M (2014) Lead gunshot pellet ingestion and tissue lead levels in wild ducks from Argentine hunting hotspots. Ecotoxicol Environ Saf 103:74–181

Ferreyra H, Beldomenico PM, Marchese K, Romano M, Caselli A, Correa AI, Uhart Marcela U (2015) Lead exposure affects health indices in free-ranging ducks in Argentina. Ecotoxicology 24(4):735–745

Finkelstein ME, Doak DF, George D, Burnett J, Brandt J, Church M, Grantham J, Smith DR (2012) Lead poisoning and the deceptive recovery of the critically endangered California condor. PNAS 109(28):11449–11454

Finley MT, Dieter P, Locke LN (1976) Lead in tissues of mallard ducks dosed with two types of lead shot. Bull Environ Contam Toxicol 16(3):261–269

Franson JC, Hansen SP, Creekmore TE, Brand CJ, Evers DC, Duerr AE, DeStefano S (2003) Lead fishing weights and other fishing tackle in selected waterbirds. Waterbirds 26(3):345–352

Fry DM, Maurer J (2003) Assessment of lead contamination sources exposing California condors. In: Species conservation and recovery program report, pp 1–85

Garcia RC, Snodgrass WR (2012) Lead toxicity and chelation therapy. Am J Health Syst Pharm 64:45–53

Godwin HA (2001) The biological chemistry of lead. Curr Opin Chem Biol 5:223–227

Heinz GH, Hoffman DJ, Sileo L, Audet DJ, LeCaptain LJ (1999) Toxicity of lead-contaminated sediment to Mallards. Arch Environ Contam Toxicol 36(3):323–333

Hoffman DJ, Franson JC, Pattee OH, Bunck CM, Murray HC (1985) Biochemical and hematological effects of lead ingestion in nestling American kestrels (Falco sparverius). Comp Biochem Phys C 80(2):431–439

Hoffman DJ, Heinz GH, Sileo L, Audet DJ, Campbell JK, LeCaptain LJ (2000a) Developmental toxicity of lead-contaminated sediment to mallard ducklings. Arch Environ Con Toxicol 39 (2):221–232

Hoffman DJ, Heinz GH, Sileo L, Audet DJ, Campbell JK, LeCaptain LJ, Obrecht HH III (2000b) Developmental toxicity of lead-contaminated sediment in Canada Geese (Branta canadensis). J Toxic Environ Health A 59(4):235–252

Hoffman DJ, Rattner BA, Burton GA, Cairns J (2003) Handbook of ecotoxicology, 2nd edn. Lewis Publishers, Boca Raton, p 1290

Hoffman DJ, Heinz GH, Audet DJ (2006) Phosphorus amendment reduces hepatic and renal oxidative stress in mallards ingesting lead-contaminated sediments. J Toxicol Environ Health A 69(11):1039–1053

Holladay JP, Nisanian M, Williams S, Tuckfield RC, Kerr R, Jarret T, Tannenbaum L, Holladay SD, Sharma A, Gogal RM Jr (2012) Dosing of adult pigeons with as little as one #9 lead pellet caused severe δ-ALAD depression, suggesting potential adverse effects in wild populations. Ecotoxicology 21(8):2331–2337

Howard DR, Penumarthy J (1979) Lead poisoning in Canada geese: a case report. Vet Hum Toxicol 21(4):243–244

Huang H, Wang Y, An Y, Tian Y, Li S, Teng X (2017) Selenium for the mitigation of toxicity induced by lead in chicken testes through regulating mRNA expressions of HSPs and seleniproteins. Environ Sci Pollut Res Int 24(16):14312–14321

Johnson MS, Pluck H, Hutton M, Moor G (1982) Accumulation and renal effects of lead in urban population of feral pigeons (*Columbia livia*). Arch Environ Contam Toxicol 11:761–767

Jones MM (1994) Design of new chelating agents for removal of intracellular toxic metals. In: Coordinating chemistry: a century of progress. American Chemical Society, Washington, pp 427–438

Jordan JS (1968) Influence of diet in lead poisoning in waterfowl. Trans NE Sect Wildl Soc 25:143–170

Kelly ME, Fitzgerald SD, Aulerich RJ, Balander RJ, Powell DC, Stickle RL, Stevens W, Cray C, Tempelman RJ, Bursian SJ (1998) Acute effects of lead, steel, tungsten, iron, and tungsten-polymer shot administered to game farm mallards. J Wildl Dis 34:673–687

Kelly TR, Grantham J, George D, Welch A, Brandt J, Burnett J, Sorenson KJ, Johnson M, Poppenga R, Moen D, Rasico J, Rivers JW, Battistone C, Johnson CK (2014) Spatiotemporal patterns and risk factors for lead exposure in endangered California condors during 15 years of reintroduction. Conserv Biol 28(6):1721–1730

Kerr R, Holladay S, Jarrett T, Selcer B, Medlrum B, Williams S, Tannenbaum L, Holladay J, Williams J, Gogal R (2010) Lead pellet retention time and associated toxicity in northern bobwhite quail (*Colinus virgianus*). Environ Toxicol Chem 29(12):2869–2874

Kirberger M, Yang JJ (2008) Structural differences between Pb^{2+} and Ca^{2+} binding sites in proteins: implications with respect to toxicity. J Inorg Biochem 102:1901–1909

Kirberger M, Wong HC, Jiang J, Yang JJ (2013) Metal toxicity and opportunistic binding of Pb^{2+} inproteins. J Inorg Biochem 125:40–49

Locke LN, Bagley GE, Irby HD (1966) Acid-fast intranuclear inclusion bodies in the kidneys of mallards red lead shot. Bull Wildl Disease Assoc 2(4):127–131

Locke LN, Kerr SM, Zoromski D (1982) Lead poisoning in common loons (*Gavia immer*). Avian Dis 26(2):392–396

Luo J, Gao Z, Wang W, Hartup BK (2016) Lead in the red-crowned cranes (*Grus japonensis*) in Zhalong wetland, northeast China: a report. Bull Environ Contam Toxicol 97(2):177–183

Martinez-Haro M, Green AJ, Mateo R (2011) Effects of lead exposure on oxidative stress biomarkers and plasma biochemistry in waterbirds in the field. Environ Res 111:530–538

Mateo R, Hoffman DJ (2001) Differences in oxidative stress between young Canada geese and mallard exposed to lead-contaminated sediment. J Toxicol Environ Health A 64:531–545

Mateo R, Cadenas R, Manez M, Guitart R (2001) Lead shot ingestion in two raptor species from Donana, Spain. Ecotoxicol Environ Saf 48(1):6–10

Mateo R, Beyer WN, Spann JW, Hoffman DJ, Ramis A (2003) Relationship between oxidative stress, pathology, and behavioral signs of lead poisoning in mallards. J Toxicol Environ Health A 66(14):1371–1389

Mateo R, Green AJ, Lefranc H, Baos R, Figuerola J (2007) Lead poisoning in wild birds from southern Spain: a comparative study of wetland areas and species affected, and trends over time. Ecotoxicol Environ Saf 66:119–126

Mauser DM, Rocke TE, Mensik JG, Brand CJ (1990) Blood lead concentrations in mallards from Delevan and Colusa National Wildlife Refuges. Calif Fish Game 73(3):137–145

McConnell CA (1967) Experimental lead poisoning of bobwhite quail and mourning doces. In: Proceedings of the Southeastern Association of Game and Fish Commissioners, New Orleans, pp 208–219

McElvaine MD, DeUngria EG, Matte TD, Copley CG, Binder S (1992) Prevelance of radiographic evidence of paint chip ingestion among children with moderate to severe lead poisoning, St. Louis, Missouri. Pediatrics 89:740–742

Milton LA (1988) Lead and lead poisoning from antiquity to modern times. Ohio J Sci 88:78–84

Needleman HL (2000) The removal of lead from gasoline: historical and personal reflections. Environ Res Sect A 84:20–35

Newth JL, Rees EC, Cromine RL, McDonald RA, Bearshop S, Pain DJ, Norton GJ, Deacon C, Hilton GM (2016) Widespread exposure to lead affects the body condition of free-living whooper swnas *Cygnus cygnus* wintering in Britain. Environ Pollut 209:60–67

Nijman P (2016) Case study: lead toxicity in an Australian harrier. In: Proceedings of the New Zealand Veterinary Nurses Association conference, pp 63–68

Nriagu JO (1983) Lead and lead poisoning in antiquity. Wiley, New York

Osickova J, Band'ouchova H, Kovacova V, Kral J, Novotny L, Ondracek K, Pohanka M, Sedlackova J, Skochova H, Vitual F, Pikula J (2014) Oxidative stress and liver damage in birds exposed to diclofenac and lead. Acta Vet Brno 83:299–304

Pain DJ (1989) Hematological parameters as predictors of blood lead and indicators of lead-poisoning in the black duck (*Anus-Rebripes*). Environ Pollut 60:67–81

Pain DJ (1991) Lead shot densities and settlement rates in Carmague marshes, France. Biol Conserv 57(3):273–286

Rabinowitz MB (1991) Toxicokinetics of Bone Lead. Environ Health Perspect 91:33–37

Redig PT, Arent LR (2008) Raptor toxicology. Vet Clin North Am Exotic Anim Pract 11 (2):261–282

Riecke TV, Conway WC, Haukos DA, Moon JA, Comer CE (2015) Baseline blood Pb concentrations in black-necked stilts on the upper Texas coast. Bull Environ Contam Toxicol 95 (4):465–469

Romano M, Ferreyra H, Ferreyroa G, Molina FV, Caselli A, Barberis I, Beldomenico P, Uhart M (2016) Lead pollution from waterfowl hunting in wetlands and rice fields in Argentina. Sci Total Environ 545–546:104–113

Ruiz S, Espin S, Rainio M, Ruuskanen S, Salminen JP, Lilley TM, Eeva T (2016) Effects of dietary lead exposure on vitamin levels in great tit nestlings – An experimental manipulation. Environ Pollut 213:688–697

Sadler RC, LeChance J, Hanna-Attisha M (2017) Social and built environmental correlates of predicted blood lead levels in the Flint water crisis. Am J Public Health 107(5):736–769

Samour JH, Naldo J (2002) Diagnosis and therapeutic management of lead toxicosis in falcons in Saudi Arabia. J Avian Med Surg 16(1):16–20

Schueuhammer AM (1997) Influence of reduced dietary calcium on the accumulation and effects of lead, cadmium, and aluminum in birds. Environ Pollut 94:337–343

Schulz JH, Millspaugh JJ, Washburn GR, Wester JT, Lanigan J III, Franson JC (2002) Spent-shot availability and ingestion on areas managed for Mourning Doves. Wildl Soc Bull 20:112–120

Shacklette HT, Boerngen JB (1984) Element concentrations in soils and other surficial materials of the conterminous United States. US Geological Survey Professional Paper, p 1270

Shimmel L, Snell K (1999) Case studies in poisoning: two eagles lead and pesticide toxicity in two challenging cases. J Wildl Rehabil 22(1):8–17

Sidor IF, Pokras MA, Major AR, Poppenga RH, Taylor KM, Miconi RM (2003) Mortality of common loons in New England, 1987 to 2000. J Wildl Dis 39(2):306–315

Snyder NFR, Snyder H (2000) The California condor: a saga of natural history and conservation. Academic Press, San Diego

St Clair CT, Baird P, Ydenberg R, Elner R, Bendell LI (2015) Trace elements in pacific dunlin (*Calidris alpina pacifica*): patterns of accumulation and concentrations in kidneys and feathers. Ecotoxicology 24(1):29–44

Strimbu K, Tavel JA (2010) What are biomarkers? Curr Opin AIDS 5(6):463–466

Sun GX, Chen Y, Liu CP, Fu J (2016) Effect of selenium against lead-induced damage on the gene expression of heat shock proteins and inflammatory cytokines in peripheral blood lymphocytes of chickens. Biol Trace Elem Res 172(2):474–480

Tchounwou PB, Yedjou CG, Patlolla AK, Sutton DJ (2012) Heavy metals toxicity and the environment. EXS 101:133–164

Trampell DW, Imerman PM, Carson TL, Kinker JA, Ensley SM (2003) Lead contamination of chicken eggs and tissues from a small farm flock. J Vet Diagn Investig 15(5):418–422

Tsipoura N, Burger J, Newhouse M, Jeitner C, Gochfeld M, Mizrahi D (2011) Lead, mercury, cadmium, chromium, and arsenic levels in eggs, feathers, and tissues of Canadian geese of the New Jersy Meadowlands. Environ Res 111(6):775–784

US Environmental Protection Agency (USEPA) (1994) Lead fishing sinkers; response to citizens' petition and proposed ban. Fed Regist 59:11122–11143

US Environmental Protection Agency (USEPA) (2005) Best management practices for lead at outdoor shooting ranges. EPA-902-B-01-001

US Geological Survey (USGS) (2016) Minneral commodity summaries 2016, pp 96–97

Vallverdu-Coll N, Lopez-Anita A, Martinez-Haro M, Ortiz-Santaliestra ME, Mateo R (2015a) Altered immune response in mallard ducklings exposed to lead through maternal transfer in the wild. Environ Pollut 205:350–356

Vallverdu-Coll N, Ortiz-Santaliestra ME, Mouget F, Vidal D, Mateo R (2015b) Sublethal Pb exposure produces season-dependent effects on immune response, oxidative balance and investment in carotenoid-based coloration in red-legged partridges. Environ Sci Technol 49 (6):3839–3850

Vallverdú-Coll N, Mougeot F, Ortiz-Santaliestra ME, Castano C, Santiago-Moreno J, Mateo R (2016) Effects if lead exposure on sperm quality and reproductive success in an avian model. Environ Sci Technol 50(22):12484–12492

Veit HP, Kendall RJ, Scanlon PF (1983) The effect of lead shot ingestion on the testes of adult ringed turtle doves (*Streptopelia risoria*). Avian Dis 27(2):442–452

Vermeulen A, Muller W, Matson KD, Tieleman BI, Berveots L, Eens M (2015) Sources of variation in innate immunity in great tit nestlings living along a metal pollution gradient: an individual-based approach. Sci Total Environ 508:297–306

Walters JR, Derrickson SR, Fry DM, Haig SM, Marzluff JM, Wunderle JM Jr (2010) Status of the California condor (*Gymnogyps californianus*) and efforts to achieve its recovery. Auk 127 (4):969–1001

Wayman GA, Tokumitsu H, Davare MA, Soderling TR (2011) Analysis of CaM-kinase signaling in cells. Cell Calcium 50:1–8

Wiemann M, Schirrmacher K, Busselberg D (1999) Interference of lead with the calcium release activated calcium flux of osteoblast-like cells. Calcif Tissue Int 65:479–485

Zheng S, Song H, Gao H, Liu C, Zhang Z, Fu J (2016) The antagonistic effect of selenium on lead-induced inflammatory factors and heat shock protein mRNA level in chicken cartilage tissue. Biol Trace Elem Res 173(1):177–184

Index

P. de Voogt (ed.), *Reviews of Environmental Contamination and Toxicology Volume 245*, Reviews of Environmental Contamination and Toxicology 245,
https://doi.org/10.1007/978-3-319-75037-8

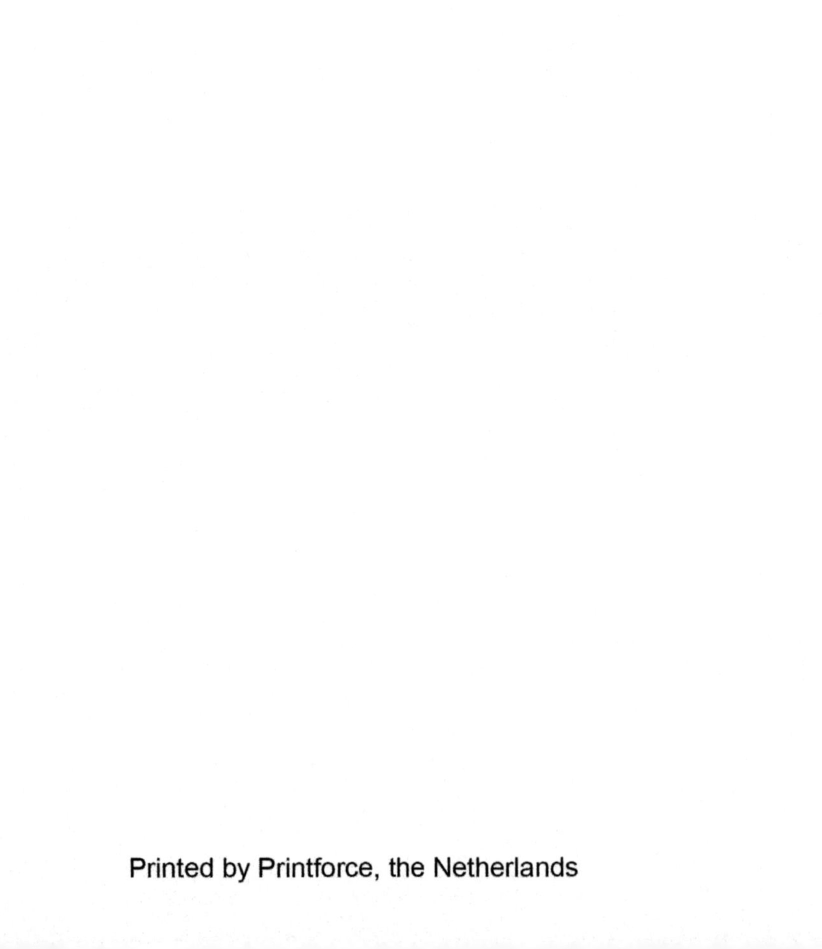

Printed by Printforce, the Netherlands